STRUCTURED QUES

MATHEMATICS

BOOK ONE

Sallie Hudson

HODDER AND STOUGHTON
LONDON SYDNEY AUCKLAND TORONTO

CONTENTS

PREFACE

This book is designed to be used with the majority of students following GCSE courses although some questions will not be suitable for the top level of GCSE. Like many of those found in GCSE papers, the questions are based on practical situations. The situation is described and then questions are asked to examine and develop the understanding of the mathematical aspects of that situation.

It is intended that the book be used as a classroom resource and as a basis for tests, homework, and pre-exam revision. Its use will provide a practical and relevant setting for common mathematical concepts.

The division of the contents under chapter headings should be seen only as a convenient guide as most real situations cannot usually be categorised in terms of their mathematical content. The questions become more difficult as one progresses through a chapter, and some of the questions at the end of a chapter may not be appropriate for some students.

I would like to thank Alastair McMurdo for checking and advising on the language level and on some of the technical material, and also for proof-reading.

Sallie Hudson

1 Carpet designs

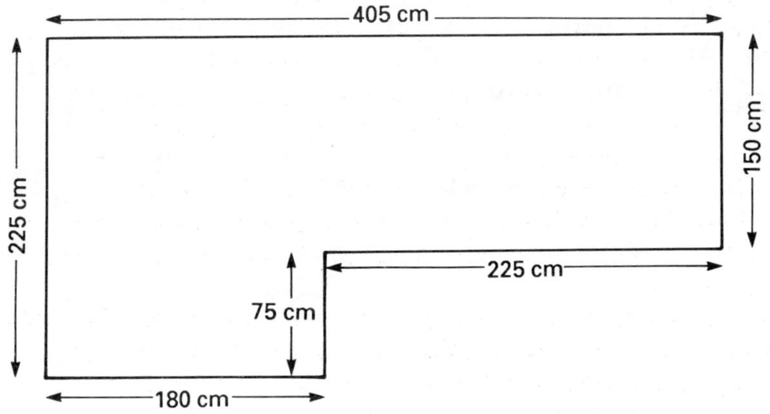

This is the plan of a room (not drawn to scale) whose floor is to be covered with carpet tiles.

(a) Make 3 identical scale drawings of this room on squared paper using a scale of 1 cm to 30 cm. *(2)*

(b) Carpet tiles are squares of carpet with side length 30 cm which are placed together to cover a floor. Assume 2 different colours of carpet tile are available. Using the same scale for the tiles, on each of your drawings create a different pattern for covering the floor. (Note that some of the tiles may have to be cut to fit the floor.) *(6)*

(c) As carpet tiles are expensive, you want to use the pattern which will cost the least. For each of your patterns calculate how many of each colour of tile you have used. *(3)*

(d) The tiles are only sold in whole packs of 10 of the same colour. Look at your answers to (c) above and work out which would be the cheapest design. *(2)*

(e) The packs cost £5.00 each. Work out the cost of the cheapest pattern. *(2)*

Total 15 marks

2 Pascal's supermarket

In many supermarkets you see pyramid-shaped displays of tinned foods. The top four layers of a display of tinned spaghetti would look like this from above:

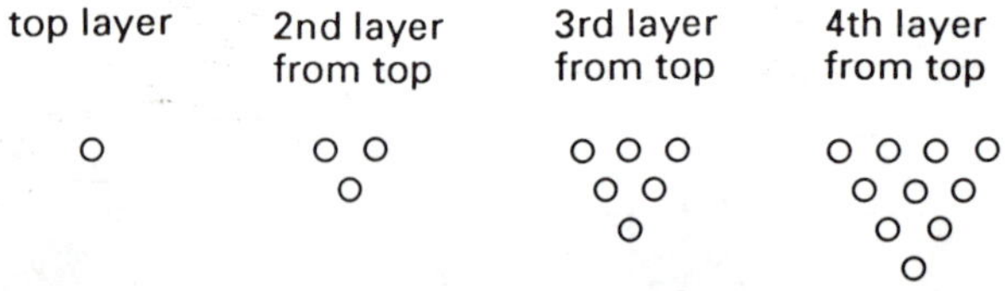

(a) If a pyramid is wanted which is to be as close to 1 m high as possible, but no more than 1 m high, and if a tin of spaghetti is 11 cm high, how many layers will be needed? *(1)*

(b) (i) Write down the number of tins in each layer above.
(ii) How many tins would be needed for the next layer down? (A drawing might help.)
(iii) Continue the pattern down to the bottom layer and describe the pattern. *(6)*

(c) From your pattern, how many tins are needed for the bottom layer of the pyramid? *(1)*

(d) Copy the following table, complete it, and continue it to the bottom layer:

layer	*number of tins in layer*	*total number of tins needed down to the layer indicated*	*height of a pyramid down to the layer indicated*
top	1	1	11 cm
2nd	3	4	22 cm
3rd	6	10	

(3)

(e) How many tins are needed altogether for the pyramid? *(1)*

(f) How many tins are needed for a pyramid 55 cm high? *(1)*

(g) How high would a pyramid of 20 tins be? *(2)*

Total 15 marks

3 Borders

A large square floor in a dance hall is to be covered in square tiles in the following pattern:

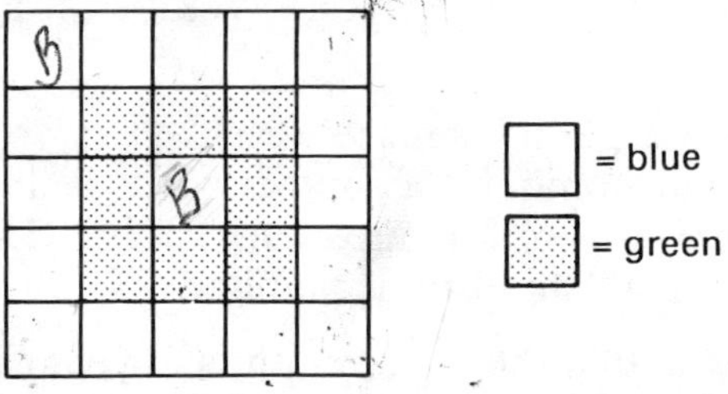

8 green tiles are needed to surround the central blue tile.

(a) How many blue tiles are needed to surround the 8 green tiles? *(1)*

(b) The dance hall owners want to expand the floor keeping the same pattern. How many green tiles would be needed to surround the present outer layer of blue tiles? *(2)*

(c) The owners want to make the hall the biggest dance floor in the world! Copy and complete the following table to show the effects of expanding the floor:

border	*number of tiles*	*colour*
1st	8	green
2nd		blue
3rd		green
4th		blue
5th		green
6th		blue
7th		green
8th		blue
9th		green
10th		blue

(3)

(d) Describe the pattern you have found. *(2)*

(e) What is the total number of each colour of tile if the floor has 10 borders? *(2)*

Total 10 marks

4 Palindromes

Palindromes are words which are the same whether written backwards or forwards, e.g. MUM, RADAR. A palindromic sentence is more difficult to find, but 'Was it a cat I saw?' is one example.

(a) Make a list of 6 palindromic words. *(2)*

(b) A palindromic number is a number which is the same written backwards or forwards, e.g. 1221, 202, 131. Make a list of 10 palindromic numbers. *(2)*

Writing dates so that the day, month, and year are written as 2 digits, e.g. 2nd May 1988 becomes 02.05.88, means that some dates become palindromic. For example, if your birthday was 27th November 1972, then you have a palindromic birthday—27.11.72.

In the same format, the first day of the first year of this century would be written 01.01.01. This is not palindromic. The only palindromic date that year would be 10.11.01.

In the second year of this century, the first day would be written 01.01.02. Again this is not palindromic. The only palindromic date that year would be 20.11.02.

In the third year of this century, the only palindromic date would be 30.11.03, and in the fourth year there would not be any palindromic date.

(c) Explain why there is no palindromic date in the fourth year of this century. *(2)*

(d) What would be the next palindromic date after the fourth year of this century? *(1)*

(e) Make a list of all the palindromic dates this century from the 1st January 1900 to 31st December 1999 using the same format as above. *(5)*

(f) How many palindromic dates are there in a century? *(1)*

(g) Describe any patterns you have found. *(2)*

Total 15 marks

5 Seeing squares

If you were asked to say how many squares were on a chessboard, you would probably answer 64 as a chessboard has 8 squares to a side.

64 is certainly the number of small squares on the board, but there are other squares of bigger size. For example, the board itself is one big square, giving us a total now of 65 squares. We could then draw a line vertically down the middle of the board and another horizontally across the middle. This would divide the board into 4 squares giving a total of 69 squares. We could combine the small squares into groups of 4 with 2 small squares to a side, so that we have a number of 2×2 squares. In the same way we could make 3×3 squares, and so on, so that we have many more squares than our original 64.

If we look at a smaller problem, we can try to find a pattern:

On a 1×1 board there is just 1 square.
On a 2×2 board there are 4 small squares plus one 2×2 square making a total of 5 squares.
On a 3×3 board there are 9 small squares plus four 2×2 squares plus one 3×3 square making a total of 14 squares.

(a) Draw diagrams to show how the totals for each of the three boards above are found. *(4)*

(b) For a 4×4 board, we can break down the problem as follows:

size of square	*number of these squares*
1×1	16
2×2	
3×3	
4×4	
Total	

Copy and complete the above table to show how many squares there are altogether on a 4×4 board. *(5)*

(c) Working in the same way, calculate how many squares there are altogether on a 5×5 board. *(2)*

(d) Describe any pattern you have found in your results. *(3)*

(e) Use your pattern to answer the following questions:

  (i) How many squares are there altogether on a 6×6 board?

  (ii) How many squares are there altogether on a 7×7 board?

  (iii) How many squares are there altogether on a chessboard? *(6)*

Total 20 marks

6 The power of money

We have 7 different coins in use in Britain at the moment—1p, 2p, 5p, 10p, 20p, 50p, £1. If you had one of each coin you could make several amounts of money depending on how many of them you used. The highest amount you could make would be £1.88 using all seven. The lowest amount you could make would be 0p by using none of the coins. Between 0p and £1.88 there are many other possible amounts.

(a) We will need more than one of each coin to make some amounts. For example, to make 4p we would need either four 1p coins, or one 2p coin and two 1p coins, or two 2p coins. Write down three amounts which cannot be made with only one of each coin. 3, 4, 6 *(2)*

(b) Finding out how many amounts can be made when there is only one of each of the seven coins is easier to do by starting with a simpler problem. Copy the following table and continue it until you get to seven coins. (Remember that you always have the choice not to use any of the coins when working out what amounts can be made.)

coins used	*number of coins used*	*amounts that can be made*	*number of amounts that can be made*
none	0	0p	1
1p	1	0p, 1p	2
1p, 2p	2	0p, 1p, 2p, 3p	4
1p, 2p, 5p			?

(Note: The 3rd column may get too difficult to finish, but by then you may have found a pattern in the 4th column which will help you to complete the rest of the table.) *(4)*

(c) Describe any patterns and rules that you have found. *(3)*

(d) How many different amounts can be made from the seven coins if none of them are used more than once? *(1)*

Total 10 marks

7 Filling carriages

A small train with open carriages is run in the Welsh mountains during the tourist season. The train has six carriages, three holding 4 people and the other three holding 6 people. Thus the maximum number of people it can take is 30:

3 carriages holding 4 people	12
3 carriages holding 6 people	18
total	30

As with most tourist attractions the costs of running the train are quite high and the train only runs at a profit when the carriages are full. If there is one spare seat, it makes neither a profit nor a loss. If there are 2 spare seats, it makes a small loss. The operators will not use carriages with more than 2 spare seats. Any number of carriages in any combination can be used. We could look at all the possible combinations in a table as follows:

number of 4-seat carriages used	*number of 6-seat carriages used*	*maximum number of people carried*
1	0	4

(a) Copy the table and complete the first two columns to show all the possible combinations of the six carriages that could be used. *(5)*

(b) Fill in the third column of the table to show the total number of passengers that could be seated for each combination of carriages. *(3)*

(c) Make a list of all numbers from 1 to 30. Mark each number which respresents full carriages from your answers at (b). *(1)*

(d) Describe any pattern you see. *(2)*

(e) Cross out those numbers on your list which would mean that the train would be running with 3 spare seats. *(2)*

(f) Look at the numbers that are left. List those which represent the numbers of people where the train makes a small loss. *(2)*

Total 15 marks

8 Keeping the heat in

'Did you know that properly installed insulation can cut heating costs by as much as 40% and will also improve the comfort of your home?'

This is the kind of question found in leaflets advertising insulation for homes. Home insulation has become very popular recently due to the amount of publicity telling us how much heat is lost through the roof, walls and windows. The heat loss for an average house which does not have insulation is distributed as follows:

60% through the walls and the roof;
10% through the windows;
15% through the floors;
15% through ventilators.

(a) Imagine you are working for a company who sell home insulation. Construct a pictogram to illustrate the above information, making it as interesting as possible so that it will catch peoples' attention. *(5)*

(b) If home insulation can really cut heating costs by 40% and if your heating costs were £100 per year, then you would save 40% of £100, i.e. £40. Work out the savings in heating costs per year for the following people:

name	*heating cost per year*
Frances Hollins	£400
Bill Toothill	£350
Margaret Jones	£220
Jean Cavanagh	£230

(4)

(c) It will cost Jean Cavanagh £380 to insulate her 2-bedroomed house. How many years will it take before the money she saves in heating costs covers the cost of insulating her home? *(2)*

(d) Margaret Jones wants to insulate her home but has very little money. She decides to insulate the loft only and to do the job herself. This costs her £88. She is advised that during the next year she will save £22 on her heating bill.

(i) How many years will it take before the money she saves on heating will cover the cost of insulating her loft?

(ii) What will Margaret Jones save on her heating bill next year as a percentage of this year's heating cost? *(4)*

Total 15 marks

9 Electricity costs

Nearly everyone has electricity in the home in Britain today. Some appliances use more electricity than others, e.g. a cooker uses more electricity in one hour than a light bulb does. How much more depends on the Watt rating. The higher the Watt rating, the more electricity used.

A 1000 watt (usually written 1000 W) appliance switched on for 1 hour uses 1 unit of electricity which costs about 6p.

(a) For 6p you can use a 1000 W appliance for 1 hour. For 6p you can use a 2000 W appliance for ½ hour. Copy and complete the table below to show how long each of the appliances can be used for 6p.

appliance	*watt rating (W)*	*number of hours (h)*	*cost (p)*
toaster	1000 W	1 h	6p
kettle	2000 W	½	6p
electric fire	4000 W	¼	6p
vacuum cleaner	500 W	2	6p
television	100 W	10	6p
light bulb	100 W	10	6p
electric whisk	200 W	5	6p

(4)

(b) Using the information above, what would it cost to use an electric fire for 2 hours? 48p *(2)*

(c) What would it cost to use a television for 5 hours a night for 4 weeks? 84p *(3)*

(d) The following advertisement is trying to persuade you to buy a television:

> 'For the price of one daily newspaper you can watch this television all evening for a whole week.'

Make up 2 advertisements like this using the information about other appliances in your table. Try to make them as tempting as possible! *(6)*

Total 15 marks

10 Cabbage patch

Winston is planning his vegetable garden. He wants to grow leeks, potatoes, onions, cabbages, lettuce, and broccoli this year. He only has a small vegetable garden, a rectangle 3 m wide by 4 m long, and he is going to plant his rows along the width of the garden, i.e. each row will be 3 m long.

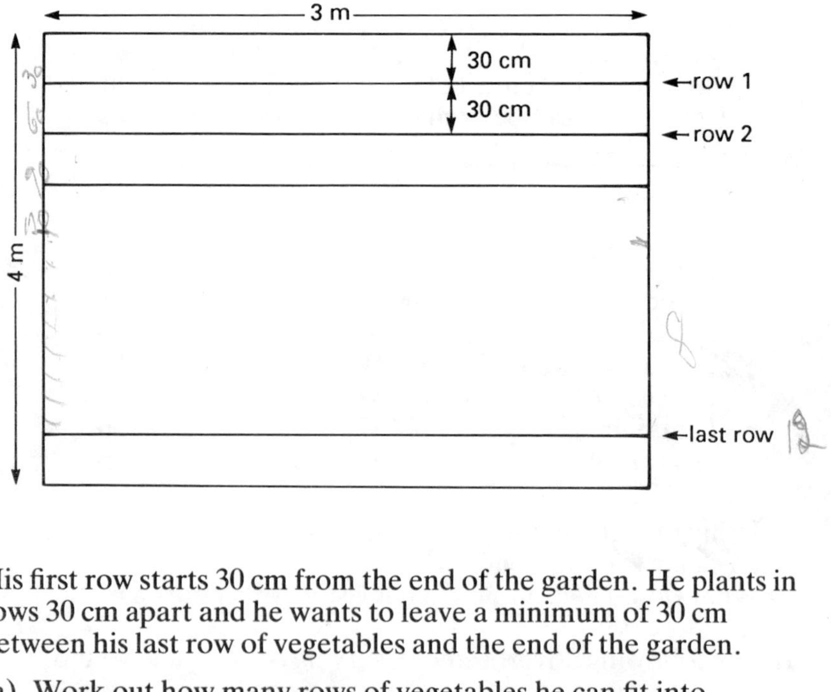

His first row starts 30 cm from the end of the garden. He plants in rows 30 cm apart and he wants to leave a minimum of 30 cm between his last row of vegetables and the end of the garden.

(a) Work out how many rows of vegetables he can fit into his plot. *(2)*

He decides to plant 3 rows of potatoes and 1 row of lettuces. The remaining rows are to be divided equally among the other vegetables.

(b) How many rows of cabbages will he plant? *(2)*

The variety of cabbage he has chosen can be sown from early March to late May. He does not want all the cabbages to be ready to pick at the same time so he decides to sow some of the seeds every 2 weeks from March 12th to May 21st inclusive.

(c) How many times will he sow cabbage seeds? *(2)*

(d) Each time he sows cabbage seeds he will only use a section of each row. He wants to sow the same number of seeds each time, so each section should be the same size. How many sections should he divide *each* cabbage row into so that he has one section for each time he sows cabbage seeds? *(1)*

(e) What will be the length in cm of each section? *(1)*

(f) The seeds are sold in packets, each one containing about 400 seeds. About how many cabbage seeds should he use each time he sows? *(1)*

(g) Winston does not want to count out the seeds as this would take too long. How else might he divide up the seeds? *(1)*

Total 10 marks

11 The cost of nothing

Kelvin got a Saturday job working in a small grocery shop in his village. The shop provided a delivery service and one of Kelvin's jobs was to work out the bills for the customers. The shop charges a 5% delivery charge on every order. Customers receive a monthly bill for all their deliveries in the previous month. If they pay within one week they are allowed a 5% discount.

The manager explained all this to Kelvin on his first day and then left him to get on with working out the bills for the previous month. However, he could not understand why he had to add 5% and then take the 5% off again. He thought he must have misunderstood the manager and decided to make up the bills without adding the 5% delivery charge or taking away the 5% discount.

(a) Make up a bill for the following order:

12 East St Mr Gordon	
2 loaves bread	at 36p per loaf
4 tins beans	at 22p per tin
5 lb potatoes	at 11p per lb
1 lb butter	at 62p per ½ lb
500 g cheese	at £3.60 per kg
6 tins soup	at 26p per tin
2 packets custard creams	at 16p per packet
Box of tea bags	at £1.44
2 lb carrots	at 44p per lb
Box of tissues	at 75p
250 g mixed nuts	at £3.60 per kg
Swiss roll	at 36p
3 yoghurts	at 20p each

(5)

(b) What is the total cost of this order? *(2)*

When Kelvin returned the next week, the manager gave him a calculator and asked him to do all the bills again. She explained that adding 5% and then taking 5% off the resulting total was not the same as doing nothing at all.

(c) Work out the 5% delivery charge on Mr Gordon's order. *(2)*

(d) What is the total charge now for this order? *(1)*

(e) Work out the 5% discount for Mr Gordon from the total in (d). *(2)*

(f) What is the total charge now with the 5% discount? *(1)*

(g) How much is the difference between what Kelvin originally charged Mr Gordon and what he should have charged by following the manager's instructions? *(1)*

(h) Another customer got the same order every month. Kelvin calculated her monthly bill as £28.60, but it should have been £28.53. By how much would this customer be overcharged by Kelvin over a whole year? *(1)*

Total 15 marks

12 A fair price to pay

Many schools have Parent-Teacher Associations (PTAs) who, from time to time, organise social events for parents and teachers. One such PTA in Bradford plans to hold a Wild West night with entertainment provided by the school drama club and some parents and teachers. The event is to take place in the school hall. The secretary of the PTA has the job of costing the event and working out the price to charge for tickets. As it is a purely social event they are not aiming to make a profit, but of course do not want to make a loss either.

Below is a list of estimated costs:

Payment to caretaker for the evening		£10.00
Decorating the hall		£40.00
Food:		
Sausages	£1.90 per kg (1 kg provides 10 portions)	
Beans	48p per large tin (1 tin provides 4 portions)	
Rolls	53p per dozen	
Coffee/tea	20p per person	

The PTA plans to cater for up to 100 people. They are printing 100 tickets (at no cost to the PTA) and the secretary needs to work out a fair price.

(a) For 100 people, how many of the following will be needed: (i) kg of sausages (ii) tins of beans (iii) dozens of rolls? *(3)*

(b) Work out the *total* cost of the evening for 100 people assuming everyone has a portion of sausages, a portion of beans, one roll, and one cup of tea or coffee each. *(3)*

(c) Calculate the cost per person to the nearest penny. *(2)*

(d) This cost assumes that all 100 tickets are sold, no food or drink is wasted, and no breakages occur. This is perhaps not being practical. How much do you think the secretary should charge? Explain your answer. *(3)*

(e) If only 75 tickets are sold at the price you have suggested at (d), how much money will be taken? *(2)*

(f) How much profit or loss will result from this given the cost at (b)? *(2)*

Total 15 marks

13 The school tuck shop

The job of a stock control clerk in a warehouse is to keep track of the value of the stock and to ensure that under-stocking and over-stocking do not occur. Running the school tuck shop also involves stock control. Below is a stock list for the 4th year tuck shop at the beginning of the week:

item	*quantity in stock*	*cost price per item*	*quantity sold*	*selling price per item*
milk choco bars	10	25p	5	26p
crisps—plain	5	16p	42	17p
—tomato	6	16p	26	17p
—cheese & onion	1	16p	32	17p
—salt & vinegar	2	16p	16	17p
sherbert fountains	8	5p	30	6p
button bubble gum	20	4p	54	5p
smudge fudge bars	12	24p	12	25p
nut & raisin pyramids	9	18p	7	20p
paradise creams	12	12p	14	13p

The first column shows the quantity of each item in stock.
The second shows how much each item costs to buy into the shop.
The third shows the quantity sold during the *previous* week.
The fourth shows how much the shop sells each item for.

Every Monday a teacher is asked to buy the stock for the week.

(a) From the stock list above, make out a list of items which you think need to be bought for this week. *(2)*

(b) As all items are only sold in boxes of 12, how many boxes of each of the items at (a) will be needed? *(3)*

(c) From the 'cost price', work out how much cash the teacher will need. *(4)*

(d) Work out the profit on each item that is sold in the shop. *(1)*

(e) From the 'quantity sold', work out the profit which the tuck shop made last week. *(3)*

(f) If there are 119 pupils in the 4th year, how many items did each pupil buy on average last week? *(2)*

Total 15 marks

14 Exchange rates and exports

Much talk goes on in the news about the value of the £ and how it affects business in Britain, but how does it affect an exporting business? Imagine a business which makes beds and currently sells them in Britain, but is about to start exporting them to Spain.

The firm's double beds cost £120 to produce and are sold in Britain at £210 each. Their single beds cost £40 to produce and are sold in Britain at £80 each.

(a) (i) How much profit is made by selling a double bed in Britain?
(ii) How much profit is made by selling a single bed in Britain? *(2)*

The firm wishes to make the same profit when it starts selling in Spain, but there are export costs to take into account. These work out at an average cost of £25 per bed.

(b) For how much will the firm have to sell double and single beds in Spain to make the same profit as it does in Britain? *(2)*

The unit of currency in Spain is the peseta.

(c) Find out the current exchange rate for pesetas. *(1)*
(d) Using this exchange rate and your answers to (b), calculate the selling prices of double and single beds in Spain. *(2)*
(e) If the value of the pound rises so that it is worth more pesetas, how will this affect the prices of the beds in Spain? *(1)*
(f) Will a rise in the value of the pound affect the number of beds sold in Spain by this firm? Explain your answer. *(1)*
(g) Will a rise in the value of the pound affect the number of beds sold in Spain by Spanish bed manufacturers? Explain your answer. *(1)*

Total 10 marks

15 Mutations

In 1850 an unusual type of moth was discovered in Manchester. It was a mutation of a species called the Peppered Moth. The normal Peppered Moth had speckled wings. The mutation had black wings. These moths lived on trees which by 1850 had become black with pollution. The black-winged moths were therefore well disguised on trees, but the speckled-winged moths could be seen easily and were eaten by birds. Thus the black-winged population increased and the speckled-winged decreased.

Suppose that twice as many black-winged offspring survived into adulthood as speckled-winged offspring. The diagram below shows the effect on the second generation (the offspring) if the first generation is made up of 4 speckled-winged moths (shown as an S) and 1 black-winged moth (shown as a B):

1st generation	2nd generation
S S	S S
B	B B
S S	S S
4 speckled-winged moths and 1 black-winged moth	4 speckled-winged moths and 2 black-winged moths

(a) Copy this diagram and continue it to show the distribution of speckled-winged and black-winged moths in the 3rd, 4th, and 5th generations. *(3)*

(b) Calculate the percentage of black-winged moths in each of the 5 generations. Write your answers underneath the appropriate generation in your diagram. *(5)*

(c) Would the speckled-winged moths ever disappear completely? Explain your answer. *(2)*

(d) How many generations would it take for the speckled-winged moth to make up less than 1% of the whole population? *(3)*

(e) If the black-winged mutation had occurred in a large unspoilt area of countryside in 1850, do you think the population would have increased as it did in Manchester? Explain your answer. *(2)*

Total 15 marks

16 Employment trends

Employment Trends, GB 1971–1984 (millions)

year	*employees in employment*				*employed labour force*	*working population*
	all male	*all fem*	*f-t fem*	*p-t fem*		
1971	13.4	8.2	5.5	2.8	24.0	24.7
1972	13.3	8.3	5.5	2.9	23.9	24.7
1973	13.5	8.7	5.5	3.2	24.5	25.1
1974	13.4	8.9	5.5	3.4	24.6	25.1
1975	13.2	9.0	5.4	3.6	24.5	25.3
1976	13.1	9.0	5.4	3.6	24.3	25.5
1977	13.1	9.1	5.4	3.6	24.3	25.6
1978	13.1	9.2	5.5	3.7	24.4	25.7
1979	13.2	9.5	5.6	3.9	24.8	26.0
1980	13.0	9.4	5.5	3.9	24.7	26.2
1981	12.3	9.1	5.3	3.8	24.8	26.1
1982	11.9	9.0	5.1	3.9	23.3	26.1
1983	11.7	8.9	5.0	3.9	23.0	26.0
1984	11.6	9.1	4.9	4.2	23.4	26.5

[fem = female, f-t = full-time, p-t = part-time.]

[Source: Employment Gazette 1985]

(a) (i) Write down the number of female part-time employees in 1971 and 1984.
(ii) What is the increase in number?
(iii) Work this out as a percentage increase. *(3)*

(b) (i) Write down the number of female full-time employees in 1971 and 1984.
(ii) What is the decrease in number?
(iii) Work this out as a percentage decrease. *(3)*

(c) (i) Write down the number of male employees in 1971 and 1984.
(ii) What is the decrease in number?
(iii) Work this out as a percentage decrease. *(3)*

(d) Look at your answers to (a), (b), and (c) above and comment on the pattern of employment between 1971 and 1984. *(2)*

(e) The last two columns of the table give the numbers actually employed and the number in the working population. In 1971, 24.0 million out of 24.7 million were working. That is an employment rate of 97%.
   (i) Calculate the employment rate in 1983 and 1984.
   (ii) Compare these percentages for 1971, 1983, and 1984 and comment on your results. *(4)*

Total 15 marks

17 A series of fractions

Going to pop concerts is an expensive business. The cost of a trip with bus fare and entrance fee included may be £20 or more. Suppose you want to go to such a concert with a total cost of £20.48, and your mother agrees to lend you the money on certain conditions. She will give you half of the cost, i.e. £10.24, now so you can put a deposit on your ticket. She will then give you half of the remaining amount tomorrow providing you help her with any work she has to do. Similarly, on the day after, she will give you half of what then remains provided you help, and so on until you finally have the £20.48.

(a) She gives you half the cost of the trip, i.e. £10.24, on Saturday, and then after helping, you receive half of the remaining amount, i.e. £5.12, on Sunday. How much will she give you on Monday, provided you help? *(1)*

(b) Continue the calculation for each day after that until you find that your mother can no longer give you any money. *(2)*

(c) How many days altogether has it taken to get this far? *(1)*

(d) (i) How much has your mother now given you?
(ii) Is this enough for the trip? *(2)*

(e) Copy the diagram below and complete each sum. Draw a number line for each sum, making sure that there is an equal distance between each number line. Mark the answer to each sum accurately at the correct position on the line with a cross.

$1 + \frac{1}{2} = 1\frac{1}{2}$ — 0 ——— 1 ——— 2

$1 + \frac{1}{2} + \frac{1}{4} = 1\frac{3}{4}$ — 0 ——— 1 ——— 2

$1 + \frac{1}{2} + \frac{1}{4} + \frac{1}{8} =$

$1 + \frac{1}{2} + \frac{1}{4} + \frac{1}{8} + \frac{1}{16} =$

$1 + \frac{1}{2} + \frac{1}{4} + \frac{1}{8} + \frac{1}{16} + \frac{1}{32} =$

$1 + \frac{1}{2} + \frac{1}{4} + \frac{1}{8} + \frac{1}{16} + \frac{1}{32} + \frac{1}{64} =$

$1 + \frac{1}{2} + \frac{1}{4} + \frac{1}{8} + \frac{1}{16} + \frac{1}{32} + \frac{1}{64} + \frac{1}{128} =$ *(6)*

(f) Explain any pattern you see in your results. *(2)*
(g) What would be the answer to the next line if you had continued with the diagram above? *(1)*
(h) If you continued, would the answer ever reach 2? *(1)*
(i) Join the crosses in your diagram. Try to make the line curve smoothly through all the crosses. What do you notice about the curve? *(2)*
(j) Now try to explain your answer to (h) above. *(2)*

Total 20 marks

18 Display boards

A school is having an open day. In the centre of the school hall the art teacher wants to put up a display of work using display boards. The display boards are 1 m wide. He is given 12 for the display. The head teacher asks him to arrange the 12 boards in a rectangle in the middle of the hall. There are several ways the teacher could combine the boards, e.g.:

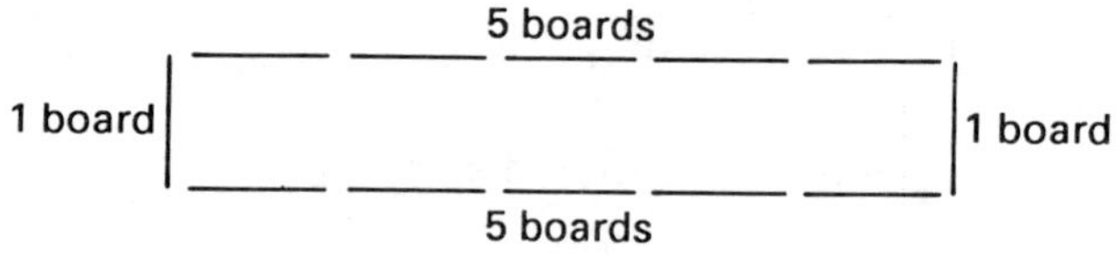

However, the head teacher wants him to arrange the boards in such a way that they take up as little floor space as possible.

(a) Draw 2 other rectangular arrangements of the 12 boards. *(2)*

(b) Work out the floor space taken up by each of the 3 arrangements. *(3)*

(c) Which one of the 3 arrangements should the art teacher use to take up as little space as possible. *(1)*

(d) The art teacher discovers that he needs 16 boards to display all the work he would like to. How many different arrangements of 16 boards would there be to choose from? *(2)*

(e) Which would take up the least space? Explain your answer. *(3)*

The head teacher decides that the art display should be set up in the school foyer after the open day. It is a large foyer and so space is not a problem. The art teacher decides to put the 16 boards in the shape of an octagon using 2 boards to each side.

(f) Using a scale of 1 cm to 1 m, make an accurate drawing of the octagon on squared paper. Estimate the area in square metres that this arrangement would occupy. *(4)*

Total 15 marks

19 Arrow marketing

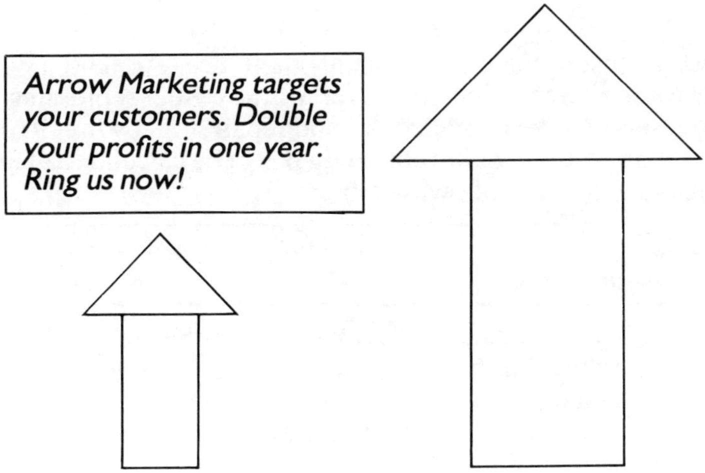

Adverts like this are often used to draw our attention.

(a) The smaller arrow is made up of a rectangle 1 cm wide and 2 cm high, and a triangle on top with base 2 cm and height 1 cm.
  (i) What is the area of the rectangle?
  (ii) What is the area of the triangle?
  (iii) What is the total area of the smaller arrow? *(3)*

(b) The larger arrow has been drawn from the smaller one by doubling the dimensions, e.g. the base of the larger arrow's triangle is twice as long as that of the smaller arrow's.
  (i) Calculate the area of the larger triangle.
  (ii) What is the total area of the larger arrow? *(5)*

(c) Look at your answers to (a)(iii) and (b)(ii). Explain why the advert could be misleading. *(2)*

(d) If the advert had said 'Triple your profits in one year', the larger arrow would have been drawn by multiplying each of the dimensions of the smaller arrow by 3.
  (i) Draw the large arrow in this way.
  (ii) Calculate the area of the large arrow.
  (iii) How many times greater is the area of the larger arrow than the smaller one? *(5)*

Total 15 marks

20 House Insurance

Most people insure their home against theft, fire, etc. The insurance will normally include cover for the cost of rebuilding the house should it be damaged beyond repair, e.g. by fire. To insure buildings, the cost of rebuilding has to be calculated. This may be done from the following table:

date built	1920–1945	1946–present
cost per square foot	£44.50	£41.50

When the date of construction is known and the area of the house in square feet has been calculated, then the table above can be used to calculate the cost of rebuilding the house.

Below is a plan of a bungalow (not drawn to scale) which was built in 1936:

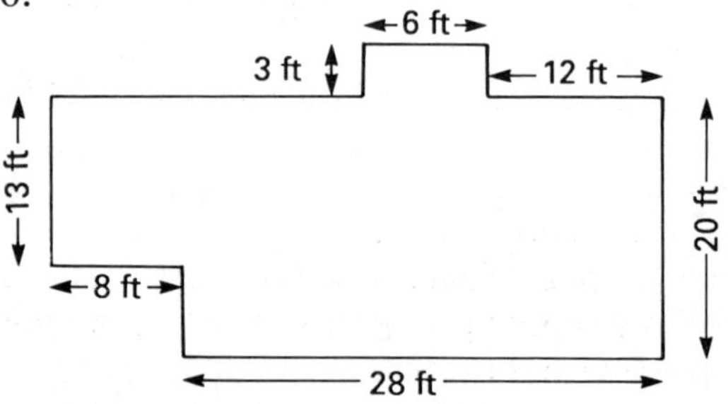

(a) Make a scale drawing of the bungalow using a scale of 1 cm to 2 ft. *(3)*

(b) Calculate the total area of the bungalow in square feet. (Hint: calculate the area of each rectangle separately first.) *(3)*

(c) Which cost per square foot in the table of rebuilding costs above do you think you should use and why? *(1)*

(d) Calculate the cost of rebuilding the bungalow. *(2)*

(e) The insurance premium (i.e. the amount you pay the insurance company each year) is quoted by the company at £2.70 for every £1000 or part of £1000 of rebuilding costs. Calculate the premium that would have to be paid to insure this bungalow. *(1)*

Total 10 marks

21 Filling glasses

A wholefood café sells apple juice by the glass. There are two sizes of glass, small and large. The small glass holds 125 ml and the large glass holds 200 ml. The café buys the apple juice in packs which contain twenty 1 litre cartons. Each pack costs £22. The café buys one pack each week.

(a) How much does 1 carton of apple juice cost the café? *(1)*

(b) How many large glasses of apple juice can be poured from 1 carton? *(1)*

(c) How much should be charged for 1 large glass of apple juice to cover its cost? *(1)*

(d) How much should be charged for 1 small glass of apple juice to cover its cost? *(1)*

(e) The café has other costs to consider and so charges more for the apple juice than the cost price calculated in (d) above. What price do you think the café would charge for:
  (i) a small glass of apple juice?
  (ii) a large glass of apple juice? *(2)*

(f) In one particular week the café sells 61 large glasses and 55 small glasses of apple juice. How much apple juice was sold altogether in litres? *(3)*

(g) Using your estimates in (e), calculate how much money the café would obtain from the sale of the apple juice in that week. *(2)*

(h) Calculate to the nearest penny the actual cost of the apple juice sold in that week. *(3)*

(i) How much profit was made from the sale of apple juice in that week? *(1)*

Total 15 marks

22 The penny farthing

The penny farthing was the popular name given to a bicycle common in Britain in the 1870s. The name came from its characteristic shape, having one very large wheel at the front and one very small wheel at the back:

The rider sat in the saddle by the front wheel which was the wheel turned by the pedals. There were no gears on penny farthings so that one complete turn of the pedals made the large wheel turn once. The distance travelled in one complete turn of the pedals depended on the size of the front wheel. The bigger the wheel, the greater the distance travelled. With the position of the saddle, taller people could manage a bigger wheel and thus could travel further in one complete turn of the pedals.

In each question below take π as 3.

(a) (i) One model had a front wheel with diameter 30 inches. How far would the bicycle travel in inches with one complete turn of the pedals?
(ii) Convert your answer to feet and inches (12 inches to 1 foot). *(2)*

(b) Another model had a front wheel with diameter 48 inches.
(i) How far would this bicycle travel in inches with one complete turn of the pedals?
(ii) Convert your answer to feet and inches. *(2)*

(c) A third model had a front wheel with diameter 64 inches.
   (i) How far would this bicycle travel in inches with one complete turn of the pedals?
   (ii) Convert your answer to feet and inches. *(2)*

(d) The small wheel of the third model had a diameter of 12 inches. How far in inches would it have travelled when it had gone round once? *(2)*

(e) Approximately how many times will the small wheel of the third model turn when the large wheel has turned once? *(1)*

(f) Draw a pair of axes. Use a scale of 1 cm to 6 inches on the horizontal axis and 1 cm to 12 inches (1 foot) on the vertical axis. Label the horizontal axis 'wheel diameter' and the vertical axis 'distance travelled in one complete turn of the wheels'.
   (i) Plot the pairs of values from your results in (a), (b), (c), and (d) above.
   (ii) Join up the points in a straight line. *(3)*

(g) (i) From your graph work out the diameter of a wheel which would travel 60 inches in one complete turn of the pedals.
   (ii) From your graph work out how far a wheel with a diameter of 15 inches would travel in one complete turn of the pedals. *(2)*

(h) Approximately how tall would someone have to be to ride the Penny Farthing whose front wheel has a diameter of 64 inches? Explain your answer. *(1)*

Total 15 marks

23 Rambling

It is very important when you go out walking in the countryside to plan your route and estimate the time your walk will take. This is to prevent the possibility of being out too long, perhaps still being out when darkness falls. Walkers use a formula as a rough guide for timing routes:

For every 4 km walked, allow 1 hour;
For every 300 m ascended vertically, allow a further 1 hour.

The following map, drawn to a scale of 2 cm to 1 km, shows one route from Horton-in-Ribblesdale up Penyghent Hill, one of the famous 'Three Peaks' of Yorkshire, and back to Horton-in-Ribblesdale:

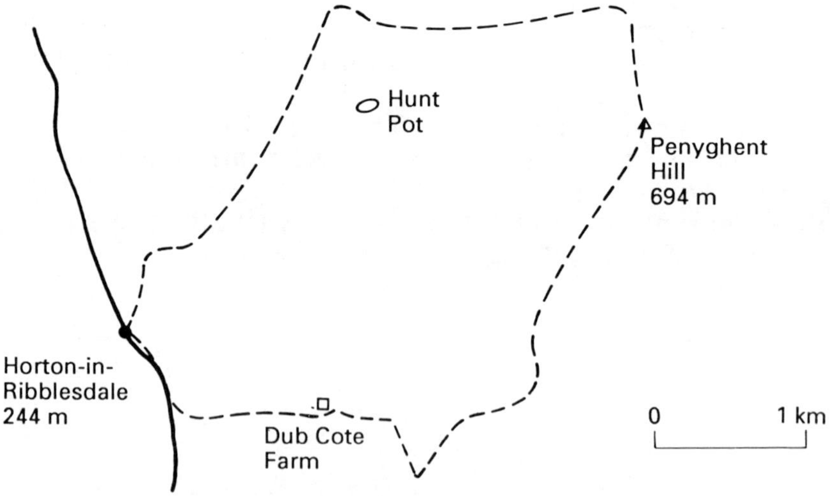

(a) By laying a piece of string along a route on a map from start to finish, you can measure the length of the route. Use this method to measure the route on the map above in centimetres. *(2)*

(b) (i) How many kilometres does this represent?
(ii) Using the rule of 1 hour per 4 km, how many hours should you allow? *(3)*

(c) (i) You also need to look at the vertical height climbed to estimate the time more accurately. Look at the height of the starting point and the height of the summit of Penyghent. What is the difference in height?

(ii) How much time should you add to your calculation for the walk to allow for this climb? *(2)*

(d) You plan to stop to eat your packed lunch part-way up. How much time would you allow for this? *(1)*

(e) When you reach the top of the hill, you may want to stop to look at the scenery, take photographs, or just recover your breath. How much time would you allow for this? *(1)*

(f) Can you think of any other reasons for stopping during your walk? How much time would you allow for these? *(2)*

(g) How much time should you allow for the walk altogether? *(2)*

(h) On the day of the walk it gets dark about 18.00. By what time should you start your walk to ensure that you finish in daylight? *(2)*

Total 15 marks

24 Hi there!

Tamasin is in town one day walking along one side of a busy pedestrian street when she sees her friend at the other side walking in the same direction. She shouts across but her friend does not hear and continues to walk straight ahead. Tamasin starts to cross the street aiming directly towards her friend. She tries to hurry, but the number of people in the street slows her down to the same speed as her friend.

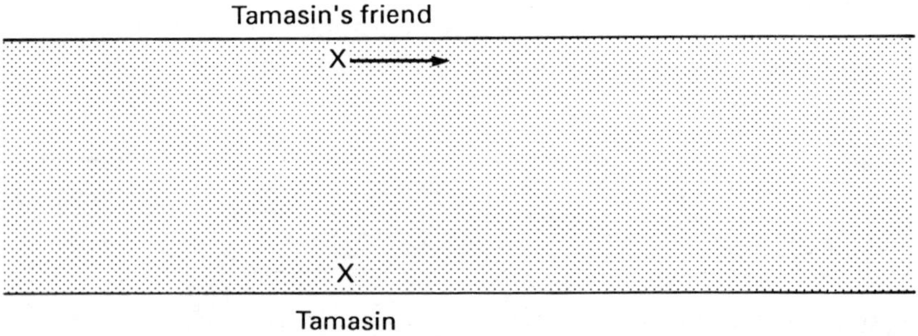

(a) Will Tamasin ever catch her friend? *(1)*

(b) Assume that for every stride Tamasin takes towards her friend, her friend takes a stride of equal length along the street. Assume the road is 10 strides wide. Copy the diagram above, using a scale of 1 cm to 1 stride. By putting a cross on the diagram for every stride the two girls take, show the movements of Tamasin and her friend. *(4)*

(c) Tamasin thinks that it may have been easier to catch her friend if the street was narrower. Using the same scale draw a diagram with a narrower street to determine if she is right. *(4)*

(d) What would Tamasin have to do to make sure she caught up with her friend? *(1)*

Total 10 marks

25 Paper sizes

Paper sizes are designed to make the most economical use of rolls of paper. Each size is given a letter and a number. Probably the most common size you use in school is A4. The diagram below is drawn to scale and shows how some of the different sizes relate to each other.

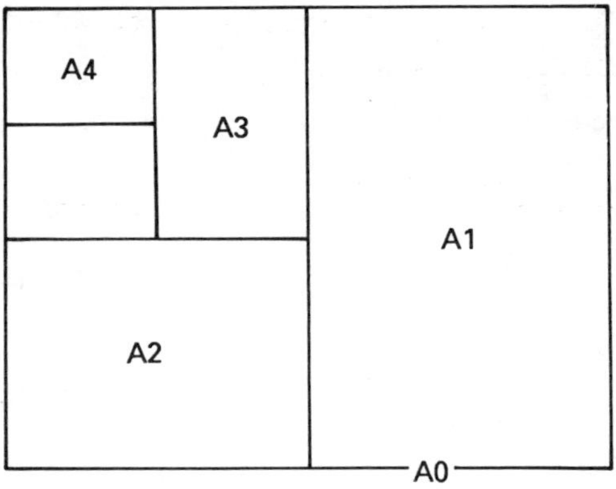

(a) Measure the sides of a piece of A4 paper and record your results. *(1)*

(b) From these measurements and using the diagram calculate the lengths and widths of A0, A1, A2, and A3 paper and record your results in a table. *(3)*

(c) The relationships between A0, A1, A2, A3, and A4 can be extended to sizes A5, A6, and A7.

(i) How many pieces of A5 do you think you could cut from a piece of A4?

(ii) What is the length and width of A5? *(2)*

(d) Complete your table for paper sizes A5, A6, and A7. *(2)*

(e) From your piece of A4, cut a piece of A5, A6, and A7 and label them. Pin these 3 pieces together through one corner in order of size. By drawing one line from the pinned corner over the 3 pieces of paper, show that A5 and A6 are enlargements of A7. *(1)*

(f) What is the scale factor of enlargement between A7 size and A5 size paper? *(1)*

Total 10 marks

26 Wedding cakes

A baker offers two different types of 3-tiered wedding cake, one circular and the other square. The diagram shows the 3 tiers of each type of cake viewed from above:

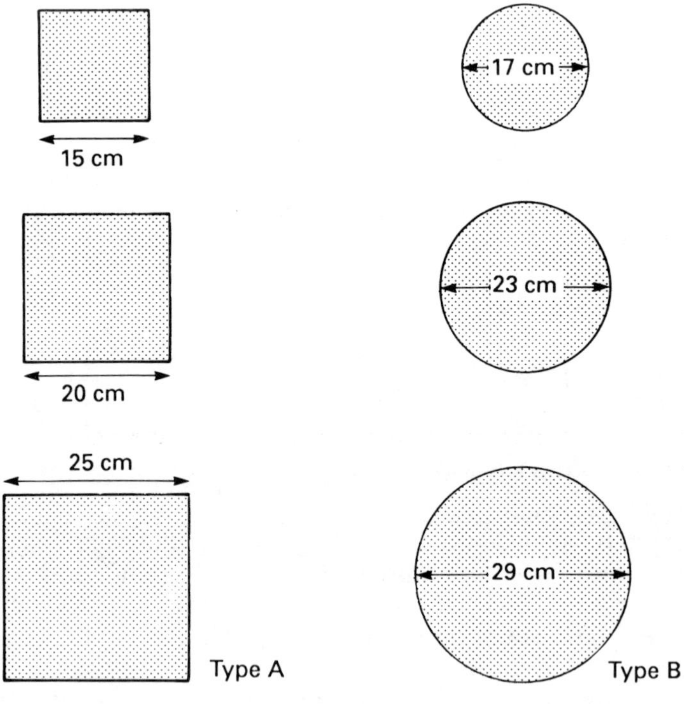

Type A:

The cake on the top tier has sides 15 cm and height 8 cm.
The cake in the middle tier has sides 20 cm and height 8 cm.
The cake on the bottom tier has sides 25 cm and height 8 cm.

(a) Work out the volume of each tier in cm^3. *(3)*
(b) What is the total volume of the Type A cake? *(1)*

Type B:

The cake on the top tier has diameter 17 cm and height 8 cm.
The cake in the middle tier has diameter 23 cm and height 8 cm.
The cake on the bottom tier has diameter 29 cm and height 8 cm.

(c) Work out the volume of each tier in cm^3. (Hint: Work out the radius of each tier first. Take π as 3.) *(3)*

(d) What is the total volume of the Type B cake? *(1)*

(e) The baker sells both cakes for the same price. Which is the better value to the customer? *(1)*

(f) The baker recommends that, in the case of the circular cake, the bottom tier be cut into 32 pieces, the middle tier into 24 pieces, and the top tier into 16 pieces. How many servings will this give for the whole cake? *(1)*

(g) Square wedding cake is usually served in pieces of equal size where the top surface of each piece measures between 3 and 4 cm wide and between 5 and 6 cm long. Draw the top surface of each of the 3 tiers of the square cake separately and show how you would cut them up for the wedding guests. *(4)*

(h) How many wedding guests can you serve out of the square cake from:
   (i) the bottom tier
   (ii) the middle tier
   (iii) the top tier? *(3)*

(i) How many wedding guests can you serve altogether from the square cake? *(1)*

(j) Which cake would you recommend to a couple getting married and why? *(2)*

Total 20 marks

27 Loading the van

The packaging department at UMAYWATE Distributors have to decide on the best way to pack a new gadget into cases. The gadget comes in a cardboard box with dimensions 20 cm wide × 30 cm long × 10 cm high. A wooden packing case is to be constructed to hold 8 of these boxes.

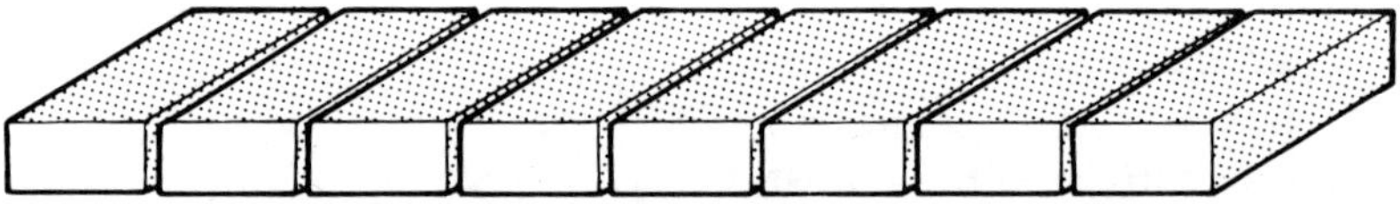

The case could be constructed so that the 8 boxes would be contained as a single row, but this would make the case very difficult to handle.

(a) Make drawings to show two different ways that the case could be constructed to hold the eight boxes. *(2)*

(b) The wood for the packing case is 2 cm thick. Work out, for both your drawings, the external dimensions of your packing cases. *(4)*

(c) What surface area of wood would be needed for each of your drawings to make the packing cases, including a lid. *(4)*

(d) Which of your two methods of packing do you think UMAYWATE would choose and why? *(1)*

(e) The delivery vans are 2.5 m high, 3 m long, and 2 m wide. Draw a diagram to show how UMAYWATE would best fill the van to get the greatest number of the type of packing case you selected at (d) into the van. *(4)*

Total 15 marks

A 'lean-to' is a type of greenhouse which is not free-standing but leans against something. I have a lean-to in my garden which rests against the garden shed. It has a rectangular base:

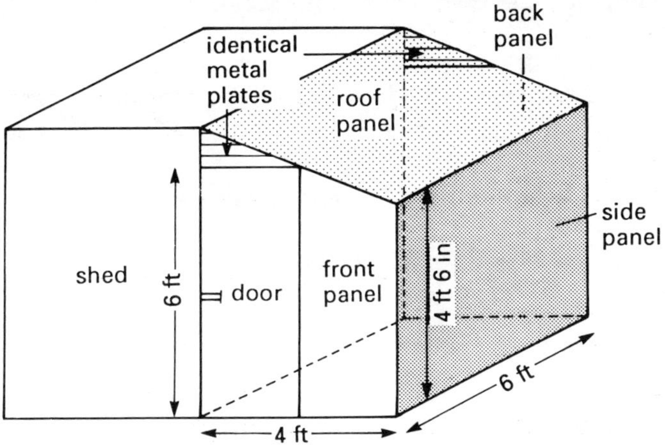

It used to be covered in polythene but last year severe winds ripped off all the polythene. I decided to re-cover it with a material called plastic glass. This was available from the local do-it-yourself shop. Plastic glass comes in rectangular sheets in a few different sizes. Sheets can be screwed together to form bigger sheets and can be cut into different shapes.

My first job was to measure those sides of the lean-to which had to be covered with plastic glass.

(a) Make separate rough sketches of each side of the lean-to to be covered with plastic glass and mark on each the dimensions given below:

Back panel:	Height:	6 ft at highest point; 4 ft 6 inches at lowest point.
	Width:	4 ft.
Front door:	Height:	6 ft;
	Width:	2 ft.
Front panel:	Height:	6 ft at highest point; 4 ft 6 inches at lowest point;
	Width:	2 ft.

Side panel:	Height:	4 ft 6 inches;
	Width:	6 ft.
Roof panel:	Width (back panel to front):	6 ft;
	Depth (side panel to shed):	5 ft.

(4)

(b) Now make accurate scale drawings of each side of the leant-to using a scale of 1 cm to 1 ft. *(6)*

(c) Work out from your scale drawing the height of the highest part of the lean-to. *(1)*

(d) The plastic glass is sold as follows:

6 ft × 2 ft	£4.59
6 ft × 2½ ft	£4.99
6 ft × 3 ft	£5.45
6 ft × 4 ft	£5.95
6 ft × 5 ft	£6.59

Work out how many of each size would be needed for each side. *(6)*

(e) Work out the total cost of the plastic glass for the new lean-to. *(3)*

Total 20 marks

29 Backwards messages

Have you noticed that some delivery and service vehicles have their name or message written in reverse? For example, if you look at the front of some fire engines you will see the word FIRE written in reverse. This is done so that drivers in front of the vehicle will be able to read the word in their driving mirrors. In other words, the letters on the vehicle are written as a mirror image.

(a) Write the word AMBULANCE as a mirror image so that it could be read correctly in a driving mirror. *(2)*

(b) (i) Make up a short advert or message that you might want written on a car.
(ii) Write it as a mirror image for drivers in front to read in their mirrors. *(3)*

(c) There are some letters of the alphabet which look the same in a mirror, e.g. A. Write down all the capital letters which look the same in a mirror. *(2)*

(d) (i) Write down 4 words which would look the same when read in a mirror.
(ii) What would be the advantage to the owner of a vehicle of using words like these in a message? *(3)*

Total 10 marks

30 Where to live

Josie and Michael are married and have both got new jobs in the north of England. Josie's is in Middlesbrough and Michael's is in Newcastle. Middlesbrough is 36 miles from Newcastle. Neither of them want to travel that distance to work every day, so they decide to live in between the two cities. As they will probably use Newcastle more than Middlesbrough for other things such as shopping trips, they decide to live no further than 18 miles from Newcastle.

(a) Mark a point on your page and label it N representing Newcastle. Using a scale of 1 cm to 3 miles, mark 6 points which are exactly 18 miles from Newcastle. *(2)*

(b) Join up these points and describe the result. *(2)*

(c) Using the same scale, mark a point M on your page representing Middlesbrough which is 36 miles from Newcastle. *(1)*

(d) Josie does not want to travel more than 24 miles to her work in Middlesbrough. Using the same scale, mark 6 points on your page which are exactly 24 miles from Middlesbrough. *(2)*

(e) Join up these points. *(1)*

(f) The two curves which join the points should overlap, enclosing an area between them.
  (i) Shade in this area.
  (ii) What does this area represent to Josie and Michael? *(2)*

Total 10 marks

31 The vegetable garden

A popular gardening radio programme suggests that if you have a very small garden and want to grow vegetables, you should grow them in a circular plot.

Experts advise vegetable growers to rotate their crops. This means that they should not grow the same type of vegetable in the same spot more than once every few years.

Don has a small garden. He decides to grow vegetables in a circular plot and divides the plot up as follows (not drawn to scale):

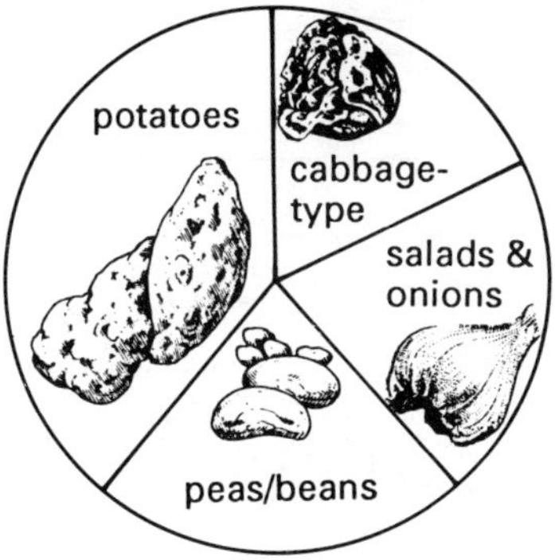

He uses a third of the plot for potatoes and divides the rest of the plot up into 3 equal parts for the rest of his vegetables, i.e. peas/beans, salads/onions, and cabbage-type vegetables.

(a) Draw an accurate plan of Don's vegetable plot, writing in the angles within each sector of the plot. *(4)*

(b) Bearing in mind the advice of the experts above, draw an accurate plan of Don's vegetable plot next year after he has rotated his vegetables so that none of the potatoes grow in the same place as this year. *(2)*

(c) For how many years can Don grow potatoes on a different piece of ground after this year? *(1)*

(d) Don's vegetable plot has an area of 4.5 m^2. What is the area of the part of the plot with
  (i) potatoes growing
  (ii) peas and beans growing? *(3)*

Total 10 marks

32 Ladders

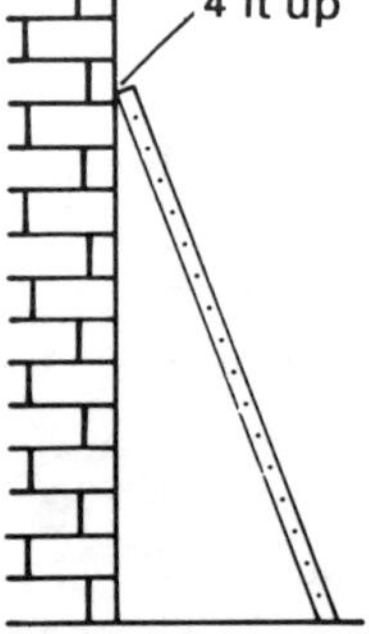

Adjustable ladders are ladders whose height can be changed. There is a useful rule for putting up adjustable ladders against a wall which says that for every 4 ft up the wall the top of the ladders have to rest, the bottom of the ladders should be placed on the ground 1 ft out from the base of the wall.

(a) How far out would you place the bottom of the adjustable ladders if you wanted to reach a window 12 ft up? *(1)*

(b) Complete this ready reckoner for using adjustable ladders:

Height up wall which ladders reach (ft)	4	8	10	12	16	20	30
Distance bottom of ladders are placed from wall (ft)	1						

(3)

(c) Make scale drawings showing the position of a ladder leaning against a wall to reach a window:
 (i) 8 ft high
 (ii) 12 ft high
 (iii) 16 ft high
Use a scale of 1 cm to 1 ft.
Measure the length of the ladder and write down this length in feet in each case. *(6)*

(d) (i) Measure the angle between the ladder and the ground and mark it on each of your drawings.
(ii) Comment on the angles you have measured. *(2)*

(e) Andrew gets the rule the wrong way round. For every 1 ft up the wall he rests the top of the ladder, he places the bottom of the ladder 4 ft away from the base of the wall.
(i) Will the angle between the ladder and the ground be bigger or smaller than in the examples you have just done?
(ii) What problems will Andrew have if he doesn't get the rule right? *(3)*

Total 15 marks

33 Who's in the picture?

If you look straight ahead, how much can you see of the things to the side of you? Look at a wall about 3 metres away from you. Imagine a line drawn from your feet to the furthest points on the wall which you can see out of the corner of your eye, as follows:

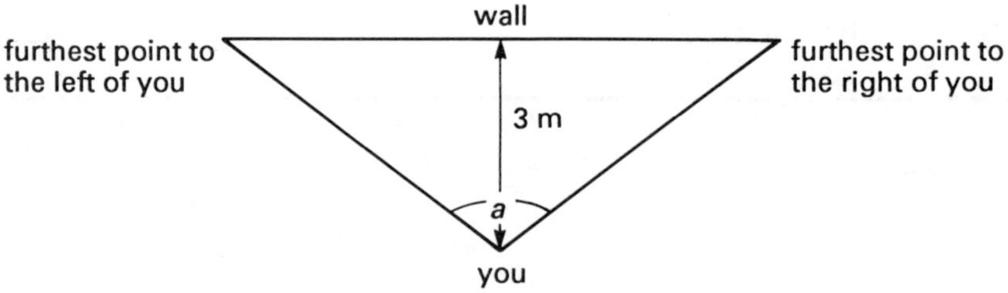

(a) What size would you estimate angle *a* to be in your case? *(1)*
(b) Make an accurate scale drawing of the diagram above making the angle equal to your answer at (a). Use a scale of 2 cm to 1 m. *(2)*
(c) Mark on your diagram a point to show where you would be standing if you moved 1.5 m closer to the wall. *(1)*
(d) The angle through which you can see things out of the corner of your eye does not change. Draw two lines from this new point to show how much of the wall you can now see out of the corner of your eye. *(1)*
(e) Describe what happens as you move towards the wall. *(2)*

Josie's Uncle Jim is about to take a wedding photograph of Josie's family who are all standing in a line facing him as follows:

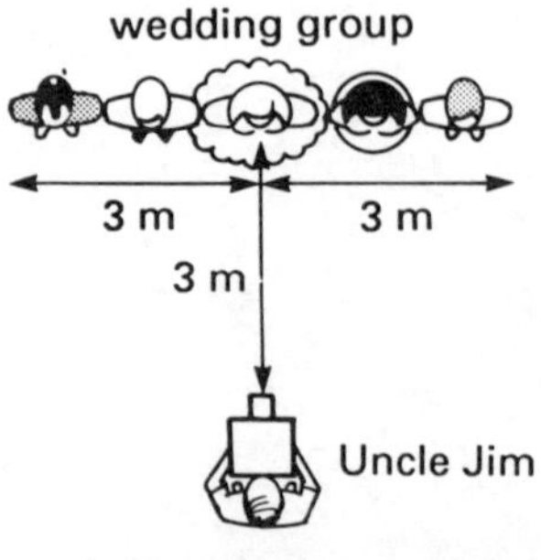

(f) Make an accurate scale drawing of the diagram using a scale of 2 cm to 1 m. *(2)*

(g) Uncle Jim's camera has a fixed lens with a viewing angle of 60°. Draw on your diagram two lines from Uncle Jim to the group to show how much of the wedding group will fit on the picture. *(1)*

(h) Should Uncle Jim move nearer to the group or further away to get all the family in the picture? *(1)*

(i) Mark on your diagram the point where Uncle Jim should stand so that he can just get everyone in the picture. (Hint: Use a protractor.) *(2)*

(j) (i) Measure how far away this point is from the middle of the group.

(ii) Convert your answer to the nearest metre. *(2)*

Total 15 marks

34 Why did the hedgehog . . . ?

Why did the hedgehog cross the road?

When travelling on country roads, you may have come across the unfortunate sight of a hedgehog which has been run over by a car. They are killed perhaps because they do not understand that roads are dangerous places and because they travel so slowly that they are likely to encounter many cars on their way across.

Hedgehogs seem usually to travel in a straight line, perpendicular to the edge of the road, taking the shortest route across. Maybe they do have some idea of road safety, or maybe most hedgehogs who used to travel diagonally across the road have been killed off as they were on the road for a longer time and thus more likely to be hit by a car.

(a) A road is 12 m wide. The hedgehog which lives by this stretch of road travels at an average speed of 2 m/min. How long does it take the hedgehog to cross the road if it travels perpendicular to the edge of the road and does not stop? *(1)*

(b) This hedgehog wants to reach a particularly interesting bit of grass on the other side of the road 5 m along from the point directly opposite. It is so anxious to reach this grass that it sets off across the road diagonally, heading straight for the point it is interested in.

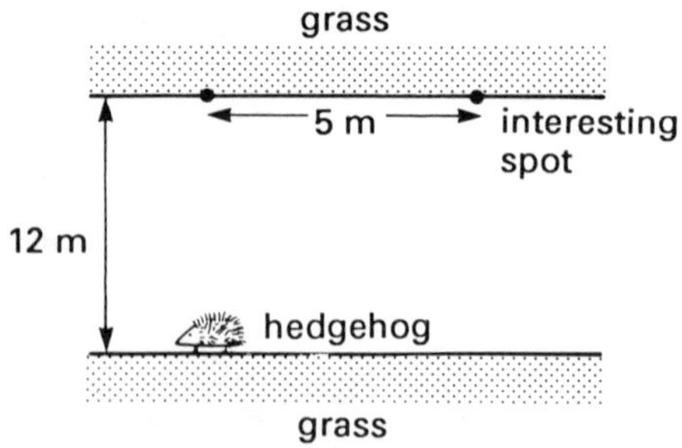

(i) How far will it travel?
(ii) How long will this take? *(3)*

(c) (i) How much longer will it be on the road travelling diagonally as at (b) rather than by taking the shortest route across the road?

(ii) What percentage increase in time on the road does this represent? *(3)*

(d) Having reached the interesting spot on the other side, it now sees an even more interesting bit of grass on the original side of the road at a point 9 m away from the point directly opposite. It sets off across the road towards this new point travelling diagonally in a straight line.

(i) How far will it travel now?

(ii) How long will this take? *(3)*

(e) (i) How much longer will it be on the road travelling diagonally as at (d) than by taking the shortest route across the road?

(ii) What percentage increase in time on the road does this represent? *(3)*

(f) Comment on your results. *(2)*

Total 15 marks

35 Working tables

Schools often have tables to work at instead of the traditional desks. In infant schools these tables are sometimes not simply rectangular, but are trapezium-shaped. When 2 of these are fitted together a regular hexagon is made:

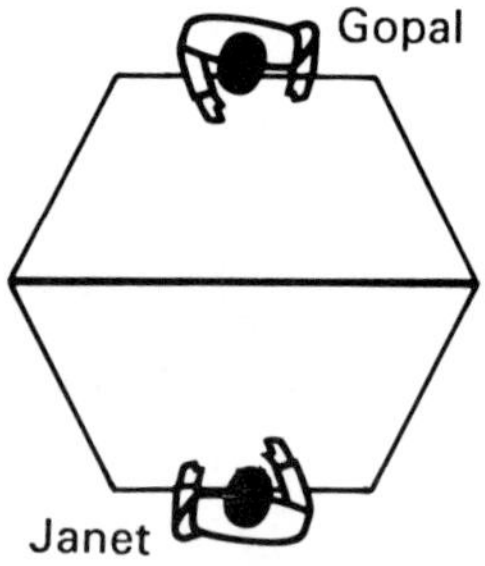

The designers of the trapezium table decided that a student needs a desk which is 50 cm wide at the front. Thus the two trapeziums fit together to make a regular hexagon with sides of 50 cm.

(a) Draw a sketch of one of these *trapezium*-shaped tables.
  (i) What is the length of the longest side of the trapezium?
  (ii) Calculate the height of the trapezium to the nearest centimetre. *(3)*

(b) Two of these tables fitted together as a hexagon can seat 6 students. Gopal and Janet are sitting opposite each other. How far apart are they? *(1)*

A school plans to furnish its resources room with as many of these trapezium tables fitted together into hexagons as it can get in the room, leaving space for the door and for people to sit and to move around. The room, shown in the diagram at the top of the next page, is rectangular with dimensions 9 m by 7 m. In one corner it has a sliding door which is 1 m wide. To determine the number of tables required a scale drawing is needed.

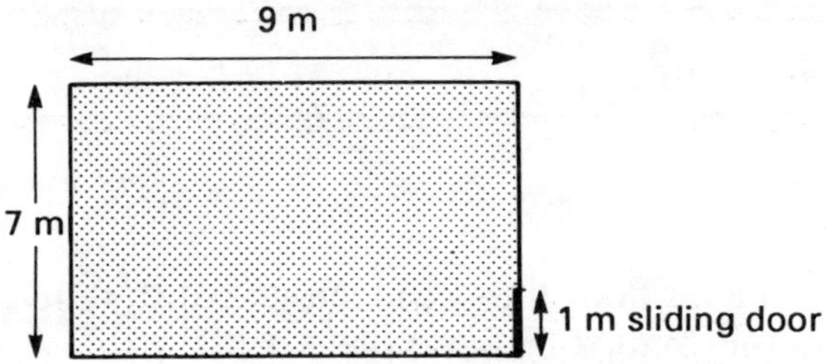

(c) Make a scale drawing of the resources room using a scale of 2 cm to 1 m. *(2)*

(d) On a separate sheet of paper make a scale drawing of two trapezium tables fitted together to make a surface which is a regular hexagon of side 50 cm, using the same scale. *(3)*

(e) Find the centre of your hexagonal surface. Using this centre draw a circle with radius 2 cm around the hexagon. *(1)*

(f) What does this circle represent? *(2)*

(g) Cut out this circular scale model. Use it as a template on your scale drawing of the room to work out how many tables you think the school should buy. Explain your answer. *(3)*

Total 15 marks

36 Fitting tiles

A firm which makes floor tiles produces two different designs. The first is simply made of square tiles:

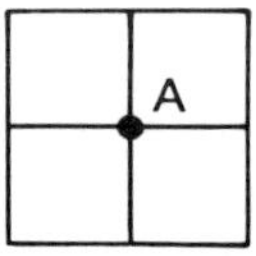

(a) How many tiles meet at point A? *(1)*
(b) What is the angle at each corner of a square? *(1)*
(c) What is the sum of the angles at A? *(1)*

The second design is made up of regular hexagons.

(d) Draw a sketch of about 8 tiles of regular hexagon shape fitting together. How many tiles meet at any one corner. *(2)*
(e) What is the angle at each corner of a regular hexagon? *(1)*
(f) A regular hexagon has 6 equal angles. What is the sum of the angles inside a regular hexagon? *(1)*
(g) An equilateral triangle has 3 angles adding up to 180°. Each angle is 60°.
A square has 4 angles adding up to 360°. Each angle is 90°.
A regular pentagon has 5 angles adding up to 540°. Each angle is 108°.
Write similar statements for a regular 6-sided figure (hexagon), a regular 7-sided figure (heptagon) and a regular 8-sided figure (octagon). *(3)*

Zoe is employed as a designer for this firm. She suggests a new floor design with 2 different-shaped tiles which can fit together without leaving any spaces in the pattern. This is a bit more expensive, but she is sure that they will sell well and the firm goes ahead with their manufacture.

(h) From your list select 2 shapes which Zoe could use. *(4)*
(i) Draw a sketch of some of the new design, adding the values of the angles in the corners. *(3)*
(j) Explain why these two shapes can be used together in a floor design. *(3)*

Total 20 marks

37 Coastguards

If a boat gets into difficulty at sea it will signal for help either by radio or, if necessary, using a flare. Coastguard stations will be alerted and they will try to estimate the boat's position at sea. They may only be able to identify a search area to which they will send a lifeboat.

For example, 2 coastguard stations situated 50 km apart may each estimate the boat's position. The first may estimate the boat's bearing to be between due north and 020°, and the second may estimate its bearing to be between 290° and 300°. These estimates will give the lifeboat an area to search for the boat.

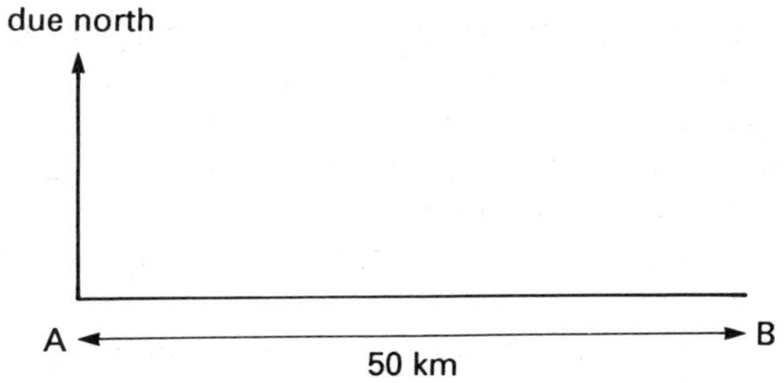

(a) Using graph paper and a scale of 1 cm to 5 km, mark the position of the coastguard stations, labelled A and B, so that B is 50 km due east of A. *(1)*

(b) From A draw a line whose bearing is due north of A. *(1)*

(c) From A draw a line whose bearing is 020° from A. *(1)*

(d) From B draw a line whose bearing is 290° from B. *(1)*

(e) From B draw a line whose bearing is 300° from B. *(1)*

(f) The section to be searched should now be clear. Estimate its area, by a suitable method, in cm^2. *(2)*

(g) How many km^2 are represented by 1 cm^2? *(1)*

(h) Estimate the area to be searched in km^2. *(1)*

(i) A lifeboat is sent from A travelling at a speed of 14 km/h. Approximately how long will it take to reach the furthest point of the search area? *(1)*

Total 10 marks

38 How far can you see?

If you stand on the seashore on a clear day and look out to sea, the horizon, the line between the sea and the sky, seems to be about 3 miles away. The following diagram illustrates this:

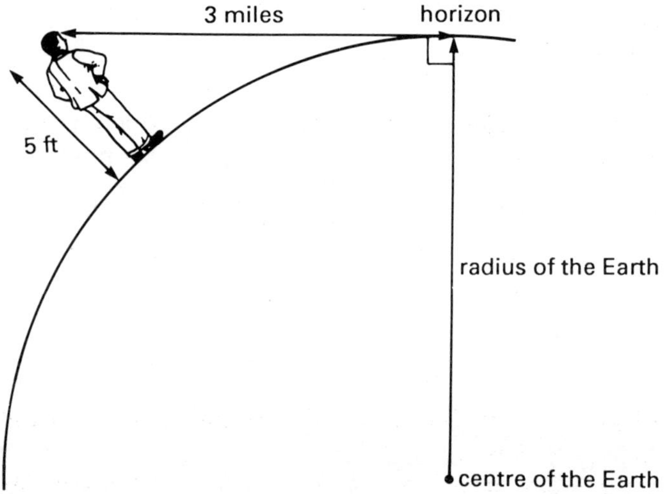

(a) Suppose that your height is 5 ft. There are 5,280 feet in a mile. Express 5 ft in miles, correct to 5 decimal places. *(2)*

(b) Using Pythagoras's Theorem and the diagram above, calculate to the nearest mile the radius of the earth. *(5)*

(c) How accurate do you think this estimate is? *(1)*

(d) The higher you are above sea level, the further away the horizon seems. Using your estimate of the radius of the Earth calculated in (b), work out the distance to the horizon if you were standing on a 300 ft cliff. *(5)*

(e) Given your ansers to (b) and (d), do you think that you would be able to see France on a clear day if you were standing on the White Cliffs of Dover? Explain your answer. *(2)*

Total 15 marks

39 Catering costs

You may have often wondered why a cup of tea in some cafés is so expensive. A tea bag, a little milk and sugar, and some hot water only cost a few pence, but you can be charged 50p or more. Some of the extra charge may go towards the rent of the café.

(a) Write down 3 other things that have to be covered by the price of the tea. *(2)*

Caterers have a formula for working out the price at which to sell their products. The formula uses a 'kitchen percentage', usually around 30–50%:

$$\text{selling price} = \frac{\text{food cost}}{\text{kitchen percentage}} \times 100$$

For example, if a cup of tea costs 10p to produce and the kitchen percentage is 40%, then:

$$\text{selling price} = \frac{10\text{p}}{40} \times 100$$
$$= 25\text{p}$$

(b) Calculate the selling prices of the following dishes using a kitchen percentage of 40%:

(i)	Omelette and chips	food cost 60p
(ii)	Bread and butter pudding	food cost 38p
(iii)	Toasted teacake	food cost 12p
(iv)	Vegetable curry and rice	food cost 72p *(4)*

The kitchen percentage is calculated from the desired 'profit percentage' as follows:

$$\text{kitchen percentage} = 100\% - \text{profit percentage}$$

(c) With a kitchen percentage of 40% what is the profit percentage? *(2)*

(d) If the caterer decides to cut the profit percentage to 50%, what will be the kitchen percentage? *(2)*

(e) Work out the selling prices of the dishes at (b) above using this new kitchen percentage. *(4)*

(f) What happens to selling prices when the profit percentage is cut? *(1)*

Total 15 marks

40 Sunrise and sunset

Sunrise and sunset change slightly each day in the time they occur. As winter approaches the days get shorter, and as summer approaches they get longer. The shortest days occur in December. On December 17th 1988 the sun rose at 08.02 and set at 15.52.

(a) For how long was there daylight on December 17th 1988? *(1)*

The longest days occur in June. On June 25th 1988 the sun rose at 04.44 and set at 21.22.

(b) For how long was there daylight on June 25th 1988? *(2)*

(c) Draw a pair of axes with months of the year on the horizontal axis starting with December and hours of daylight on the vertical axis. Sketch a rough graph to show what happens to the number of daylight hours through the year. *(4)*

(d) In 1988 British Summer Time began on 27th March. This meant that the clocks had to be put forward one hour on that date. The sun rose at 05.49 on 26th March. What time would you estimate the sun to have risen on 28th March. *(1)*

(e) From your graph, at approximately what time would you have expected the sun to set on March 28th 1988? *(2)*

Total 10 marks

41 Mirror frames

Sita helps in a workshop making mirror frames. The mirrors are all rectangular and come in various sizes. Sita's job is to calculate the amount of wood needed for the frames. She has noticed that, when measured, a few mirrors have the same perimeter although they are of different shape.

Below is a list of the 8 different sizes of mirror for which frames have to be made:

length (cm)	*width (cm)*	*length (cm)*	*width (cm)*
20	50	35	25
30	30	35	55
30	40	40	20
30	60	40	50

(a) Calculate the perimeters of the mirrors listed above. *(4)*

(b) Make a list of the dimensions of those mirrors with a perimeter of 120 cm. *(1)*

(c) Draw a pair of axes. Label the horizontal axis 'length in cm' and the vertical axis 'width in cm'. Use a scale of 2 cm to 5 cm on each axis and plot the points in your list. *(4)*

(d) Join up the points by a straight line and label the line 'perimeter 120 cm'. *(1)*

(e) Write down from your graph the dimensions of two rectangular mirrors not on Sita's list which have a perimeter of 120 cm. *(2)*

(f) Plot all the other points from Sita's list on your graph. Join the points with the same perimeters by a straight line and label these lines in the same way as at (d). *(5)*

(g) The workshop introduces a new size of mirror measuring 35 cm long by 35 cm wide. Show on your graph how you find the perimeter of this mirror. *(1)*

(h) Draw a line on your graph which you think represents mirrors with a perimeter of 130 cm and explain why it does. *(2)*

Total 20 marks

42 Temperatures

town	max temp (°)	
	C	F
Birmingham	13	55
Bristol	13	55
Buxton	11	52
Carlisle	14	57
Leeds	14	57
London	15	59
Manchester	14	57
Newcastle	13	55
Norwich	15	59
Nottingham	14	57
Plymouth	14	57
Ross-on-Wye	13	55

The above table shows maximum temperatures during an October day at various locations around England. The temperatures are given in degrees Celsius (C) and degrees Fahrenheit (F).

(a) On graph paper draw a pair of axes. Use a scale of 1 cm to 10° on each axis. Mark Celsius along the horizontal axis and Fahrenheit along the vertical axis. *(2)*

(b) (i) Water freezes at 0°C, 32°F. Plot this pair of values onto your graph.
(ii) Plot the pairs of values from the table onto your graph. *(4)*

(c) Draw a straight line through the points you have plotted and continue it. *(1)*

(d) Water boils at 100°C. Express this in °F from your graph. *(1)*

(e) On a hot day in summer a temperature of 77°F may be reached in this country. Express this in °C from your graph. *(1)*

(f) Make up a table like the one above for a day in June, giving the temperatures in Celsius and Fahrenheit. *(6)*

Total 15 marks

43 Shoe sizes

Have you ever wondered where shoe sizes come from? Here is a table showing some children's shoe sizes and their equivalent shoe lengths in inches:

shoe size (children's)	0	3	6	9	12
length of shoe (in)	4	5	6	7	8

(a) Write down any patterns you can see in the table. *(2)*

(b) Children's sizes start at size 0 and stop at size 13. The next size up is adult's size 1. Continue the table, using the same interval of 3 between shoe sizes, for adult shoe sizes finishing at size 11. *(2)*

(c) On graph paper draw a pair of axes. Use a scale of 1 cm to one shoe size on the horizontal axis and 1 cm to 1 inch on the vertical axis. *(2)*

(d) Draw a graph from your tables. *(1)*

(e) From your graph estimate the length of a baby's shoe marked 'size 1½'. *(1)*

(f) Write down your shoe size. Look at your graph and decide roughly how long your foot *actually* is. *(1)*

(g) If someone has feet 10½ inches long, what shoe size would that person take? *(1)*

Total 10 marks

44 Steep hill ahead

You may have seen road signs warning of a steep hill ahead. The form of these has been changed in recent years. The old form of warning was shown as a gradient. For example, a 1 in 5 downhill gradient was shown as:

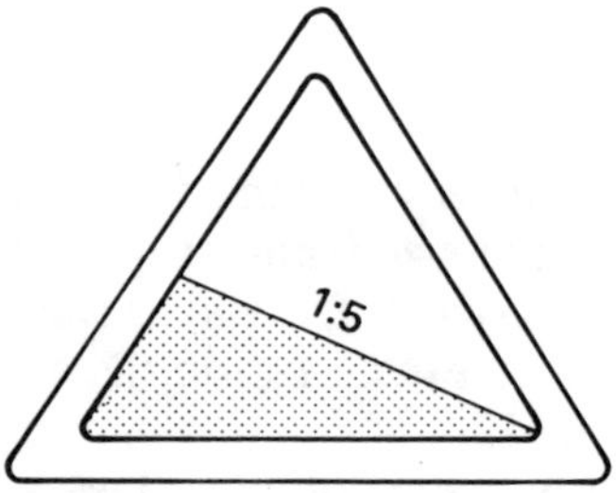

The new form is shown as a percentage. Thus a 1 in 5 downhill gradient would be shown as:

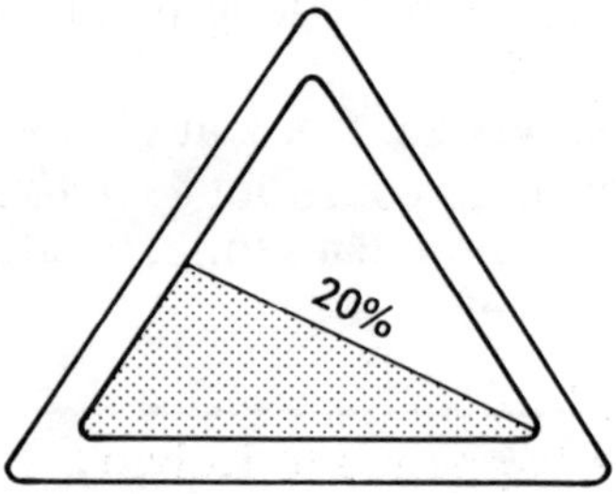

(a) Write down 5 gradients in the old form. *(2)*
(b) Convert these gradients to percentages. *(5)*
(c) Draw 5 road signs showing the above gradients in the new form. *(3)*
(d) How accurate do you think the percentages need to be? Explain your answer. *(2)*
(e) Draw an accurate diagram to show what a road with a 1 : 1 gradient would look like. *(2)*
(f) Write a 1 : 1 gradient as a percentage. *(1)*

Total 15 marks

45 The big race

At the final of the Egg and Spoon race at the local fête, there are three competitors. They have to run 100 m to the winning tape, each balancing an egg on a spoon.

The following rough sketch illustrates what happened:

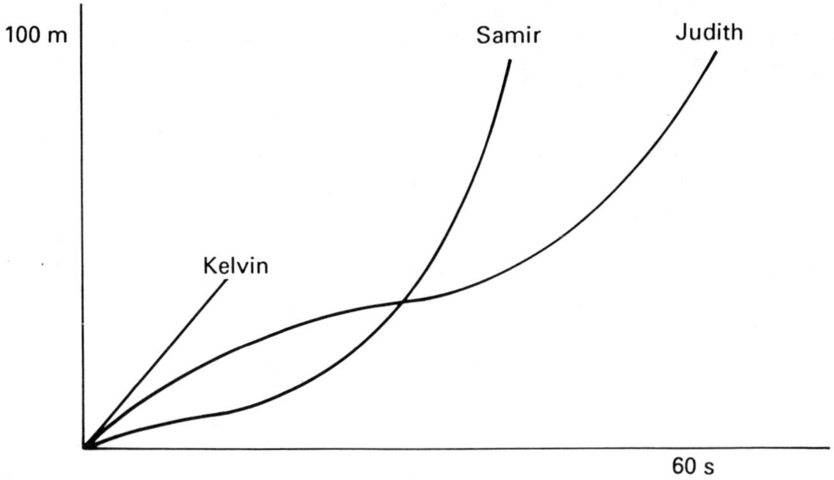

(a) Who won the race? *(1)*

(b) What do you think happened to Kelvin? *(1)*

(c) Imagine that you are the race commentator. Describe the race as accurately as you can from start to finish. *(6)*

(d) Suppose Samir had dropped his egg after 50 m but it did not break. He would have stopped, picked it up, and continued the race. Draw a similar sketch to illustrate Samir's progress in the race if this had happened. *(2)*

Total 10 marks

The Highway Code recommends safe distances which should be left between cars in motion. It suggests that for every mile per hour (miles/h) at which a car is travelling, there should be a gap of 1 metre between it and the vehicle in front. For example, when travelling at 40 miles/h, a gap of 40 m should be left.

In Holland the guideline is different as they refer to speeds in kilometres per hour (km/h). The recommendation there is that a gap in metres should be left equal to half of the speed in km/h. For example, when travelling at 50 km/h, a gap of 25 m should be left.

(a) Copy and complete the following table. (To convert miles/h to km/h divide the miles/h by 5 and then multiply the result by 8.)

miles/h	*recommended gap in UK (m)*	*km/h*	*recommended gap in Holland (m)*
20	20	32	16
30			
40			
50			
60			
70			

(5)

(b) Comment on the difference between the recommended gaps in the UK and those in Holland. *(2)*

The Highway Code gives the following figures for stopping distances at various speeds. The stopping distance is made up of a thinking distance and a braking distance, giving an overall distance for stopping in metres:

miles/h	*thinking (m)*	*braking (m)*	*overall (m)*
20	6	6	12
30	9	14	
40	12	24	
50	15	38	
60	18	55	
70	21	75	

(c) Copy and complete the 'overall' column of this table. *(1)*
(d) Describe the pattern in the figures for thinking distance. *(1)*
(e) What would be the thinking distance at 80 miles/h? *(1)*
(f) Draw a pair of axes. Using a scale of 1 cm to 10 miles/h on the x-axis and 2 cm to 10 m on the y-axis, plot the pairs of values for miles/h and overall stopping distance from your completed Highway Code table at (c). Join the points with a smooth curve. *(5)*
(g) What does the shape of your curve tell you about stopping distances and speed? *(2)*
(h) Use your graph to tell you the overall stopping distance at:
   (i) 45 miles/h;
   (ii) 75 miles/h. *(2)*
(i) In each case at (h) above, estimate how much of that distance is for thinking and how much is for braking. *(1)*

Total 20 marks

47 Taking close-ups

Most cameras cannot focus on an object closer than about 1 metre. One way of overcoming this is to use an extension tube which fits between the lens and the camera body. In doing so the amount of light reaching the film in the camera body is reduced. It is therefore necessary to allow more light to reach the film. We say that the film has to be 'exposed' to the light for longer.

A 50 mm lens has a lens length of 50 mm, i.e. the lens is 50 mm from the film in the camera body. An extension tube of 20 mm would create a new lens length of 70 mm.

The relationship between the new lens length and the exposure time necessary is:

$$\text{exposure time in seconds (s)} = \frac{(\text{new lens length in mm})^2}{2500}$$

new lens length (mm)	*exposure time (s)*
50 (no ext. tube)	1
60 (10 mm ext. tube)	
70 (20 mm ext. tube)	
80 (30 mm ext. tube)	
90 (40 mm ext. tube)	
100 (50 mm ext. tube)	
110 (60 mm ext. tube)	
120 (70 mm ext. tube)	
130 (80 mm ext. tube)	
140 (90 mm ext. tube)	
150 (100 mm ext. tube)	

(a) Copy and complete the above table using the formula. Round your answers to the nearest $\frac{1}{10}$ of a second. *(3)*

(b) On graph paper draw a suitable set of axes with lens length on the horizontal axis and exposure time on the vertical axis. *(1)*

(c) Draw the graph of the relationship between exposure time and new lens length. *(3)*

(d) From your graph estimate the exposure time needed if the lens was extended to 125 mm. *(1)*

(e) Describe your graph and explain its shape. *(2)*

Total 10 marks

48 Chewing gum

Next time you have a stick of chewing gum consider the amount of wrapping that goes into a packet of 7 sticks. Each stick is separately wrapped in silver foil and then wrapped in paper. All 7 sticks are then wrapped in foil again.

For manufacturers producing thousands of packets of gum, all this wrapping is very expensive. It is important, therefore, that the sticks of gum are wrapped in the most economical way.

Consider the following choices facing a manufacturer's Package Design department which is looking at ways of cutting packaging costs:

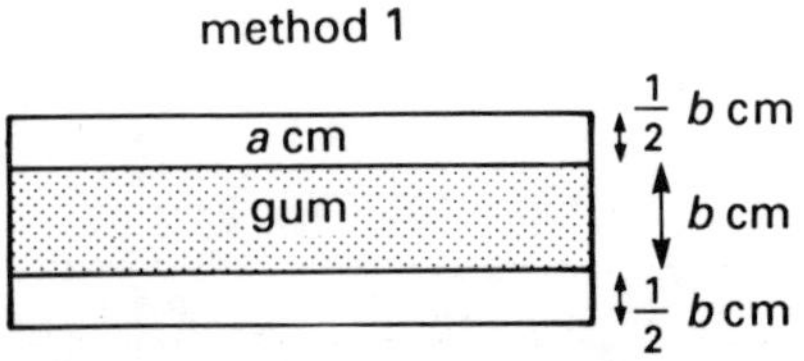

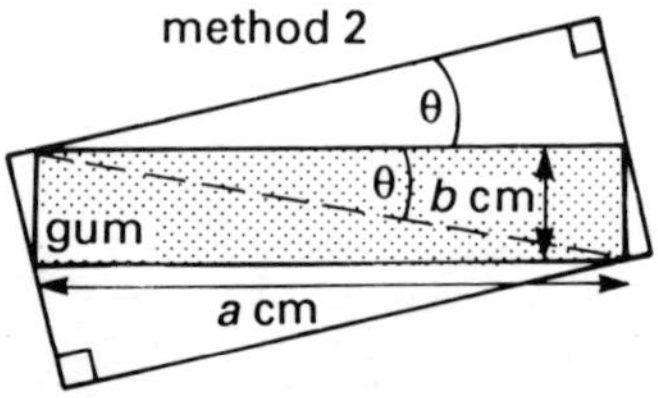

(a) Let the stick of gum have dimensions *a* cm by *b* cm. Work out an expression in terms of *a* and *b* for the area of foil needed to cover the chewing gum without any overlap in method 1. *(2)*

(b) Work out an expression in terms of *a*, *b*, and θ for the lengths of the sides of the rectangle of foil needed to cover the chewing gum using method 2. *(4)*

(c) The sticks of gum are 2 cm wide and 8 cm long. (Their thicknesses can be ignored.) Calculate the area of foil needed to cover the stick of gum without overlap using method 1. *(1)*

(d) Angle θ is the angle between the diagonal and the longest side of the chewing gum. Calculate angle θ to 1 decimal place. *(1)*

(e) Calculate the sides of the rectangle of foil used in method 2 to 2 decimal places. *(4)*

(f) Calculate the area of foil used in method 2 to 1 decimal place. *(1)*

(g) One designer suggests that the sticks be made square using the same volume of gum as before for each stick. (The thickness would remain the same.) What would be the dimensions of the square of gum? *(1)*

(h) Calculate the area of foil needed to wrap the square of gum using method 1. *(1)*

(i) Using the dimensions calculated in (e), work out angle θ in method 2 for the square of gum. *(1)*

(j) Calculate the area of foil needed for the square of gum using method 2. *(3)*

(k) There are now four alternatives facing the Package Design department. Which one would you choose and why? *(1)*

Total 20 marks

49 Double your money

Spike is always being teased by his brother Joey. On one occasion Joey told Spike that he would double Spike's money every time that he did Joey a favour.

Joey found out how much money Spike had. He then promised that for every favour Spike did he would double Spike's money if Spike then gave him back 8p. Spike thought this sounded good and agreed.

Joey first asked Spike to tidy Joey's bedroom. Spike did this, Joey doubled his money, and Spike gave Joey 8p.

Then Joey asked Spike for some of his sweets. Spike gave him some, Joey doubled his money, and Spike gave Joey 8p.

Soon after Joey asked Spike to make him a cup of cocoa. Spike did this and Joey doubled his money. Spike now only had 8p left and gave it to Joey, leaving Spike with nothing.

Spike was so cross that he told their mother who stopped both their pocket money for a week.

(a) Let the money with which Spike started be called x. Write down an expression for the money Spike had after the first favour *before* he gave 8p back to Joey. *(2)*

(b) Write down an expression for the money Spike had after the first favour *after* he gave 8p back to Joey. *(2)*

(c) Write down an expression for the money Spike had after the second favour *before* he gave 8p back to Joey. *(2)*

(d) Write down an expression for the money Spike had after the second favour *after* he gave 8p back to Joey. *(2)*

(e) Write down an expression for the money Spike had after the third favour *after* he gave 8p back to Joey. *(2)*

(f) After the third favour Spike had no money left so your expression at (e) must be equal to zero. Use this equation to find out how much Spike had at the start. *(3)*

(g) What is the smallest sum of money that Spike would have to start with so that he would not lose? *(2)*

Total 15 marks

50 Cars

The following data shows the number of people to every car in different parts of the world in 1984–85:

USA	1.8	Soviet Union	24
Australia	1.8	Africa (excl. S. Africa)	112
Western Europe	2.9	India	515
Japan	4.3	China	2022

[Source: World Watch Institute]

(a) Draw a pictogram to illustrate this data in an interesting way. *(4)*
(b) What conclusions can you draw from the data? *(3)*

The following data shows the number of cars per kilometre of road in different parts of the world in 1982–86:

Britain	62	Japan	43
USA	27	Australia	10.6
Ethiopia	1.6	Tunisia	15
India	1.1		

[Source: British Road Federation]

(c) Construct a pictogram to illustrate this data. *(4)*
(d) What conclusions can you draw from this data? *(4)*

Total 15 marks

51 Household goods

The following table shows the percentage of households in Britain possessing various household goods in 1985:

type of goods	%
refrigerator	95
deep freezer	66
washing machine	82
tumble drier	34
dishwasher	6
telephone	82
television—colour	86
—black and white only	11
video	32
home computer	13

[Source: General Household Survey 1985]

(a) What percentage of households had a washing machine? *(1)*
(b) What percentage of households had a dishwasher? *(1)*
(c) Which of the above items was possessed by the most households? *(1)*
(d) Re-order the table so that the item possessed by the most households is at the top and the item possessed by the fewest households is at the bottom. *(2)*
(e) Why do you think that 11% of households possessed a black and white television only? *(2)*
(f) Draw up a new version of the same table for this year, estimating the percentages for each item this year. Explain why you have made any differences between the 1985 figures and your new figures. *(3)*

Total 10 marks

52 Which party?

The following table gives the results of the general elections in Britain from 1922 to 1987. The columns show the number of seats won by each of the parties.

year	*Conservative*	*Labour*	*Liberal*	*other*	*total number of seats*
1922	344	142	115	14	615
1923	258	191	158	8	615
1924	412	151	40	12	615
1929	260	287	59	9	615
1931	521	52	37	5	615
1935	429	154	21	11	615
1945	213	393	12	22	640
1950	299	315	9	2	625
1951	321	295	6	3	625
1955	345	277	6	2	630
1959	365	258	6	1	630
1964	304	317	9	0	630
1966	253	364	12	1	630
1970	330	288	6	6	630
1974 (Feb)	297	301	14	23	635
1974 (Oct)	277	319	13	26	635
1979	339	269	11	16	635
1983	397	209	23*	21	650
1987	375	229	22*	24	650

* Including Social Democratic Party

(a) Make a list of the election years and beside each write down which party you think took power by having the highest number of seats. *(3)*

Sometimes a party wins a general election but does not have an absolute majority. This means that if all of the opposition vote together against the government, the government would lose the vote.

(b) Go through your list and mark the years where the winning party did not have an absolute majority. *(3)*

(c) What is the longest period that a party has remained in power since the end of the Second World War (1945)? *(2)*

(d) (i) How many times have the Conservative Party won the general election between 1922 and 1987?

(ii) How many times have the Labour Party won the general election between 1922 and 1987?

(iii) How many times have the Liberal Party won the general election between 1922 and 1987? *(3)*

(e) Draw a bar chart to illustrate your results in (d). *(4)*

Total 15 marks

53 Water consumption

The average person in Britain is said to use 30 gallons of water per day. Of these, 24 gallons are used for flushing toilets, washing, and bathing, 3 gallons are used for washing dishes, 2 gallons are used for washing cars and watering gardens, and 1 gallon is used for drinking and cooking.

(a) Construct a pie chart to illustrate these facts. *(4)*

In the African country of Zaïre water is much more difficult to obtain. The average person there is said to consume 18 pints per day. That is a little over 2 gallons.

(b) Decide how the average person in Zaïre might divide the 18 pints between the various functions of water. *(2)*
(c) Construct a pie chart to illustrate your results for Zaïre. *(4)*
(d) Explain any differences between your pie charts for Britain and Zaïre. *(2)*
(e) A person in Zaïre has to manage for all their needs on about the same amount of water as a person in Britain does just for washing cars and watering gardens. Draw a graph or pictogram to illustrate this comparison. *(3)*

Total 15 marks

54 Donkey

The card game called Donkey has a special pack of cards. The pack consists of pairs of picture cards plus one card showing a picture of a donkey. Each pair shows a different animal, both cards of each pair showing the same animal. There are 16 different animals, apart from the donkey, in the pack.

(a) How many cards are there in the pack? *(1)*

Jackie, Erin, and Lela are playing Donkey and have dealt out all the cards.

(b) How many cards do they each have? *(1)*
(c) What is the probability that Erin has the donkey card? *(1)*

The point of the game is to make up pairs of animals which are then discarded, the loser being the one who is left with the donkey. Players attempt to make up pairs by picking a card in turn from the previous player.

(d) Jackie has just taken her turn and has 4 cards left, including the donkey. It is now Lela's turn. What is the probability of Lela picking the donkey from Jackie's hand? *(1)*

Imran and Philip arrive and all five decide to start a new game of Donkey. They decide that it would be fairer to remove some cards so that everybody can start with the same number of cards.

(e) How many pairs of animal cards need to be removed? *(1)*
(f) The remaining cards are shared out equally. What is the probability of Philip getting the donkey card? *(1)*
(g) Philip starts the game. He already has the donkey card and he pulls a horse card from Erin's hand. This gives him a pair of horses which he discards. How many cards does he have left? *(1)*
(h) What is the probability that Lela, whose turn is next, will take the donkey card from Philip's hand? *(1)*
(i) Philip loses the game in the end. Imran says it is all a matter of skill. Philip says it is all chance. Whom do you agree with and why? *(2)*

Total 10 marks

55 Where to stay

A survey was conducted in 1979 to study the type of accommodation in which visitors from abroad stayed when spending their main holiday in this country. It also looked separately at where visitors stayed if their holiday was additional to their main holiday. The results, expressed as a percentage of the total number of visitors studied, were as follows:

category of accommodation	*main %*	*additional %*
licensed hotel/motel	18	17
unlicensed hotel/guest house	9	9
holiday camp	8	3
caravan	18	15
rented accommodation	15	11
camping	6	4
friends/relatives	20	37
other	6	4
total	100	100

[Source: British National Travel Survey 1979]

(a) Construct two separate pie charts to illustrate the two sets of data. *(8)*

(b) Which category of accommodation is used by most visitors for both their main and additional holiday? *(1)*

(c) One category of accommodation is used considerably more for the additional holiday than the main holiday.
   (i) Which category is it?
   (ii) Why do you think this should be? *(3)*

(d) What percentage of visitors spend their main holiday in either a holiday camp, caravan, or tent? *(1)*

(e) what percentage of visitors spend their additional holiday in either a holiday camp, caravan, or tent? *(1)*

(f) How would you explain the difference between your answers to (d) and (e) above? *(1)*

Total 15 marks

56 Dental health

For the purpose of some statistical studies on the population of Britain, the population is divided into groups according to type of employment. These groups could be specified as follows:

Group 1 Professional, employers, and managers
Group 2 Non-manual workers
Group 3 Skilled manual workers
Group 4 Semi-skilled and unskilled manual workers.

A survey on teeth was carried out in 1985 to find out the percentage of the population with no natural teeth left. The results were divided into categories according to age range and type of employment as follows:

	percentage with no natural teeth			
age range	*group 1*	*group 2*	*group 3*	*group 4*
16–24	0	0	0	0
25–34	1	1	2	3
35–44	3	4	7	14
45–54	10	14	23	34
55–64	22	34	48	58
65–74	48	53	66	71
75 and over	71	71	82	87

[Source: General Household Survey 1985]

(a) What percentage of the 35–44 age range in group 4 have no natural teeth? *(1)*
(b) Construct a bar chart for each group showing the percentage in each age range that have no natural teeth. *(8)*
(c) What happens to the percentage of people with no natural teeth in each group as age increases? *(2)*
(d) Look at the 45–54 age range. Which group has:
 (i) the highest percentage with no natural teeth?
 (ii) the smallest percentage with no natural teeth? *(2)*
(e) Write down two possible reasons for the difference in the percentages in your answers to (d)? *(2)*

Total 15 marks

57 Matrix milk

The milkman in my area has a particularly difficult job for three reasons. Firstly, there are three different types of milk. Secondly, all the houses in my street are divided into 3 separate flats. Thirdly, many of his customers, including me, do not have the same order every day. For instance, my order is one pint of milk on Mondays, Wednesdays, and Fridays and none on the other days.

I asked him how he kept a record of his deliveries and he said he used a matrix system which he had learned at school which is shown below (f = full milk, ½ = semi-skimmed, s = skimmed):

house number 6 week ending 22/10/88

$$\begin{array}{l} \\ \text{flat 1} \\ \text{flat 2} \\ \text{flat 3} \end{array} \begin{array}{c} \begin{array}{ccc} f & \tfrac{1}{2} & s \end{array} \\ \begin{pmatrix} 0 & 1 & 0 \\ 2 & 0 & 1 \\ 0 & 1 & 1 \end{pmatrix} \\ \text{Monday} \end{array} \begin{array}{c} \begin{array}{ccc} f & \tfrac{1}{2} & s \end{array} \\ \begin{pmatrix} 0 & 0 & 0 \\ 2 & 0 & 1 \\ 1 & 1 & 1 \end{pmatrix} \\ \text{Tuesday} \end{array} \begin{array}{c} \begin{array}{ccc} f & \tfrac{1}{2} & s \end{array} \\ \begin{pmatrix} 0 & 1 & 0 \\ 2 & 0 & 1 \\ 1 & 2 & 0 \end{pmatrix} \\ \text{Wednesday} \end{array}$$

$$\begin{array}{l} \\ \text{flat 1} \\ \text{flat 2} \\ \text{flat 3} \end{array} \begin{array}{c} \begin{array}{ccc} f & \tfrac{1}{2} & s \end{array} \\ \begin{pmatrix} 0 & 0 & 0 \\ 2 & 0 & 1 \\ 0 & 2 & 1 \end{pmatrix} \\ \text{Thursday} \end{array} \begin{array}{c} \begin{array}{ccc} f & \tfrac{1}{2} & s \end{array} \\ \begin{pmatrix} 0 & 1 & 0 \\ 2 & 0 & 1 \\ 2 & 0 & 0 \end{pmatrix} \\ \text{Friday} \end{array} \begin{array}{c} \begin{array}{ccc} f & \tfrac{1}{2} & s \end{array} \\ \begin{pmatrix} 0 & 0 & 0 \\ 3 & 1 & 2 \\ 1 & 2 & 1 \end{pmatrix} \\ \text{Saturday} \end{array}$$

Each week he fills in a similar sheet for every delivery.

(a) I live in house number 6. Which flat do I occupy and what type of milk do I order? *(2)*

(b) How many pints of milk does flat 2 order on a Saturday? *(1)*

The milkman explained that at the end of the week he completed another matrix to show the total orders.

(c) Complete the matrix below to show the total orders for house number 6 during week ending October 22 1988:

$$\begin{array}{l} \\ \text{flat 1} \\ \text{flat 2} \\ \text{flat 3} \end{array} \begin{array}{c} \begin{array}{ccc} f & \tfrac{1}{2} & s \end{array} \\ \begin{pmatrix} 0 & & \\ & & 7 \\ & & \end{pmatrix} \end{array}$$

(4)

(d) He sold his full and semi-skimmed milk at 28p per pint, and his skimmed milk at 27p per pint. Complete the matrices below to calculate the milk bill for each customer in house number 6 during week ending October 22 1988:

total orders		*price per pint*		*bill*	
	f ½ s				
flat 1 flat 2 flat 3	$\begin{pmatrix} & & \\ & & \\ & & \end{pmatrix}$	full semi skimmed	$\begin{pmatrix} \\ \\ \end{pmatrix}$ =	flat 1 flat 2 flat 3	$\begin{pmatrix} \\ \\ \end{pmatrix}$

(4)

(e) From 23 October 1988 his price for full milk went up by 2p per pint and his prices for skimmed and semi-skimmed by 1p. The total orders matrix for house number 6 during week ending 29 October 1988 was as follows:

	f	½	s
flat 1	0	3	0
flat 2	12	3	6
flat 3	4	9	3

Use the new milk prices to calculate the milk bills for each customer for that week. *(4)*

Total 15 marks

58 Better and better?

The following table shows the highest qualifications achieved by school-leavers. It compares the years 1975/6 and 1985/6, and boys with girls.

	boys		*girls*	
percentage with:	*1975/6*	*1985/6*	*1975/6*	*1985/6*
2 or more A levels/3 or more H grades[1]	14.3	14.9	12.1	14.2
1 A level/1 or 2 H grades	3.5	3.6	4.0	4.3
5 or more O levels/grades[2] (A–C grades)	7.2	10.0	9.4	11.9
1 to 4 O levels/grades (A–C grades[3])	23.9	24.4	27.0	28.7
1 or more O levels/grades (D or E grades) *or* CSE grades 2–5	29.9	34.0	28.4	30.9
No GCE/SCE[4] or CSE grades	21.2	13.2	19.1	10.0

Notes:

[1] H grades are the A level equivalent in Scotland.

[2] O levels are called O grades in Scotland.

[3] Includes CSE grade 1.

[4] SCE—Scottish Certificate of Education.

[Source: Department of Education and Science]

For example, a female leaving school in 1975/6 with 4 CSEs, grades between 2 and 5, would be counted among the 28.4% in the third column of figures in the table.

(a) A glance at the table suggests that standards are getting better. Would you agree with this? Explain your answer. *(3)*

(b) Compare the percentages of boys with girls at each level of achievement in 1985/6. What conclusions can you draw? *(2)*

(c) What percentage of boys left school in 1975/6 with a minimum achievement of 1 GCE O level (grade A–C)? Compare this with 1985/6? *(3)*

(d) What percentage of girls left school in 1985/6 with a minimum achievement of 1 CSE (grade 2–5)? *(2)*

(e) The new GCSE exam has been designed to enable students to achieve better results. Draw up a table based on the one above for this school year replacing O levels/grades and CSEs with GCSE grades. Estimate the percentage of boys and girls for each level of achievement. *(3)*

(f) Comment on the differences between the percentages in your table for this year and the percentages in the table for 1985/6. *(2)*

Total 15 marks

59 The probability debate

On a recent visit to a friend's house I overheard the following conversation between her teenage children Adjay and Shanaz.

'I'll toss a coin to see which of us takes the dog out,' said Adjay. 'Heads you take it out, tails I will.'

'OK,' agreed Shanaz.

Adjay tossed the coin and it landed tails up.

'It's tails, so you take him,' said Shanaz.

'Best of three, then,' said Adjay.

'OK.' Adjay tossed the coin again. 'It's tails. I win again so you take him,' said Shanaz.

'Best of five?' asked Adjay. Shanaz agreed. (I don't think she really minded taking the dog out but would get out of it if she could.)

'It's tails again. That's three in a row. I win,' said Shanaz.

Adjay was getting annoyed by now and said, 'Best of seven, and if I lose this time I'll take the dog out tomorrow as well.'

Adjay lost again. Shanaz had got four tails in a row. 'It's not fair,' shouted Adjay. 'You had so little chance of getting another tail.'

'I had exactly the same chance as I had at the start,' replied Shanaz. At this point Adjay told her she was talking rubbish and went off sulking.

The following questions should help you decide who was correct.

(a) What is the probability of getting a tail when tossing a coin? *(1)*

(b) Before starting to toss a coin, what is the probability of getting 2 tails in a row? *(2)*

(c) Before starting to toss a coin, what is the probability of getting 4 tails in a row, as Shanaz did? *(3)*

(d) Shanaz had already won 3 times and thought that the chance of getting another tail was just the same as when she got the first. Was she correct? Explain your answer. *(4)*

Total 10 marks

60 Time Share

You may have heard of Time Share. It is a holiday scheme where instead of buying a holiday home outright or renting a holiday cottage each year, you buy a holiday home for one week of the year for a number of years.

One Time Share company offers the following prices for their holiday cottages in Devon:

month	*week number*	*price (£)*	*month*	*week number*	*price (£)*
January	1	4000	July	27	9000
	2	4000		28	9000
	3	4000		29	9000
	4	4000		30	9000
February	5	4000	August	31	9500
	6	4000		32	9500
	7	5000		33	9500
	8	5000		34	9500
March	9	4500		35	9500
	10	4500	September	36	9000
	11	5000		37	8500
	12	5000		38	8500
	13	6000		39	8500
April	14	6000	October	40	8000
	15	6000		41	7500
	16	6000		42	7500
	17	6000		43	8000
May	18	7000		44	7000
	19	7000	November	45	6000
	20	7000		46	6000
	21	7000		47	5000
	22	8500		48	5000
June	23	8000	December	closed	
	24	8000			
	25	8500			
	26	8500			

(a) How much does it cost to buy the 18th week of the year? *(1)*
(b) How much does it cost to buy the 9th week of the year? *(1)*

(c) Explain why you think there is a difference in your answers to (a) and (b). *(2)*

(d) (i) Why do you think week 43 costs £8000 when week 42 only costs £7500 and week 44 only costs £7000?

(ii) Which are the cheapest weeks in the year and why? *(4)*

If you buy week 27 for £9000 from the scheme shown here, you are entitled to use the cottage for that week for the next 70 years. To rent a similar self-catering cottage in Devon in 1989 costs £250 for 1 week.

(e) How much would you spend in the next 70 years if you paid £250 per year? *(2)*

(f) Comparing the figures for Time Share and for a rented cottage, Time Share looks like a bargain. Write down two disadvantages of Time Share. *(2)*

(g) This Time Share company charges £100 for maintenance per year for each week bought. How much will this add per person to the cost over 70 years? *(1)*

(h) If the Time Share company sells all 48 weeks, what will be the value of the sales in one year? *(2)*

Total 15 marks

61 The Gregorian Calendar

It takes just less than 365¼ days for the Earth to travel round the Sun. It's that quarter day which causes us to have leap years. Back in the 16th century they had a leap year *every* four years but this was found to be too much as the calendar was getting behind the seasons.

In 1582 Pope Gregory decided that to relieve the problem he would remove 10 days from the calendar. (This caused a great deal of worry at the time as people thought that they were losing 10 days of their lives!)

He also decided that, to stop the whole thing happening again, all years exactly divisible by 4 would be leap years *except* years ending in 00.

(a) Make a list of all the years from 1582 to 5000 that end in 00. *(2)*

However this still did not quite solve the problem and he decided that all years divisible by *400* would remain leap years. With this arrangement nobody would have to worry about the calendar getting out of step with the seasons until the year 5000!

(b) Put a box round all of those years in your list which are leap years because they are divisible by 400. *(2)*

(c) All the years left unboxed in your list would have been leap years if Pope Gregory hadn't made the new rule. In 1582 the calendar was 13 days behind the seasons. How many days would we have been behind by now without Gregory's rule? *(2)*

(d) How many days behind would we have been by the year 5000? *(2)*

(e) What problems can you imagine this would have caused? *(3)*

(f) For how many years has someone born on 29th February 1964 lived? *(2)*

(g) How many *real* birthdays will they have had by now; i.e. how many times will they have been able to celebrate on 29th February? *(2)*

Total 15 marks

62 Travelling to work

Terry travels to work on a moped. His average speed when travelling on the moped is 40 km/h. On the way to work he always stops to buy a newspaper at a shop 600 m from home. This stop takes him 5 minutes. After the shop there are two sets of traffic lights and a roundabout. The first set of lights are 600 m from the shop and the second set are 300 m after the first set. The roundabout is 200 m further on from the second set of lights after which it is 500 m to work.

(a) How many seconds does it take Terry to travel 100 m on his moped? *(2)*
(b) How long does it take him to reach the paper shop from home, to the nearest minute? *(2)*
(c) If he leaves the paper shop at 08.15, what time did he leave home? *(2)*

On one particular Tuesday morning Terry had to wait for 30 seconds at the first set of traffic lights. The second set were green but roadworks at the roundabout delayed him for 10 minutes.

(d) On that Tuesday morning how long did it take him to reach work from the paper shop, to the nearest minute? *(4)*
(e) What time did he arrive at work? *(1)*
(f) Construct a distance-time graph to represent Terry's journey on that particular Tuesday morning. *(5)*

On the following Monday Terry found that the roadworks had finished and there was no traffic jam. He also got straight through both sets of traffic lights without having to stop.

(g) How long did it take him to get to work from home, to the nearest minute? *(4)*

Total 20 marks

63 Exam time

Redwood School has exams in February for the 4th year. There are 192 students in the 4th year doing a variety of subjects, some of which involve a written exam. These take place in the school hall which can hold 100 students. When there are more than 100 students sitting an exam, 3 adjacent classrooms are used, each holding 20 students. Each head of department is asked to give details of that department's exams. The head of 4th year turns this information into a table as follows:

subject	*level*	*length*	*number of students*
Maths	I	1 hour	140
Maths	II	1½ hours	154
Maths	III	2 hours	52
English	I	1 hour	103
English	II	1½ hours	160
English	III	2 hours	89
Geography		2 hours	80
History		2 hours	77
French		1½ hours	30
Biology		1½ hours	33
Science		1½ hours	100
Chemistry		2 hours	27
Physics		1½ hours	25

The following requirements have to be taken into consideration:

(1) The exams can take place from 09.00 to 12.30 and from 13.30 to 15.30.
(2) No-one taking Science is also taking Biology.
(3) No-one taking Physics is also taking French.
(4) No-one taking Chemistry is also taking English paper I.
(5) No-one taking Geography is also taking History.
(6) These written exams have to be fitted into the first three days of the week to leave time on Thursday and Friday for the practical exams.

Copy and complete the exam timetable, showing how the head

of 4th year might organise the exams to meet all the above requirements:

day	*morning*	*hall*	*rm 1*	*rm 2*	*rm 3*	*afternoon*	*hall*	*rm 1*	*rm 2*	*rm 3*
Mon	Maths I 09.00–10.00	√	√	√						
	Maths II 10.30–12.00	√	√	√	√					
Tue										
Wed										

Total 10 marks

64 Extending the season

A local netball winter league has 7 teams. Each team plays every other team once during the season, playing at a local sports hall on Thursday evenings. The hall has 2 netball courts which are booked for 1 hour. Each game lasts 30 minutes.

(a) How many matches does each team play in a season? *(1)*
(b) How many matches are played altogether in a season? *(3)*
(c) For how many weeks is the sports hall required to complete all the matches? *(2)*

The league decide to extend the season to allow the teams to play each other twice.

(d) How many matches will each team now play in a season? *(2)*
(e) How many matches will be played altogether? *(2)*
(f) For how many *more* weeks will the hall be required? *(2)*
(g) A summer league with 14 teams is also run. Each team plays every other team once. How many matches are played altogether in the summer season? *(3)*

Total 15 marks

65 Time and tides

At the coast the sea rises and falls or, to put it the more usual way, the tide comes in and goes out. The point where the tide comes furthest in is called high tide and the point where it is furthest out is called low tide. Tides are controlled by the Moon.

Each day during a lunar month (the time taken by the Moon to orbit the Earth—29½ days) the times of high tide and low tide vary. There are usually two tides per day, so there are two times for high tide and two times for low tide in one day. The height of the tides at various ports are given daily in some newspapers. They are usually measured in metres and are the heights of the water above a fixed point in the harbour at high tide and low tide.

The difference in height between high tide and low tide is called the range. For example, if, at a particular port on a particular day, high tide was measured at 5.3 m and low tide at 2.4 m, then the range would be 5.3 m − 2.4 m = 2.9 m.

There is a mathematical formula for estimating the height of the tide at any time, but sailors use a quick method called the rule of twelfths:

During the 1st hour after high tide the water falls $\frac{1}{12}$ of the range.
During the 2nd hour after high tide the water falls a further $\frac{2}{12}$ of the range.
During the 3rd hour after high tide the water falls a further $\frac{3}{12}$ of the range.
During the 4th hour after high tide the water falls a further $\frac{3}{12}$ of the range.
During the 5th hour after high tide the water falls a further $\frac{2}{12}$ of the range.
During the 6th hour after high tide the water falls a further $\frac{1}{12}$ of the range to low tide.

After low tide the tide rises again in the same way.

If high tide was 5.3 m and the range was 2.9 m then the height of the tide 1 hour after high tide would be found thus:

During the 1st hour the tide would fall $\frac{1}{12}$ of 2.9 m = 0.24 m (to 2 decimal places)
The height of the tide after 1 hour would then be:

5.3 m − 0.24 m = 5.06 m by the rule of twelfths.

(a) What would be the height of the tide 3 hours after high tide using the rule of twelfths? *(3)*

On 22nd October 1988 the following tide information was given for Hull's Albert Dock:

tide	*time*	*height*
high	17.10	8.1 m
low	23.26	3.0 m

(b) Calculate the range at Albert Dock for 22nd October. *(1)*

(c) Copy and complete the following table for Albert Dock by using the rule of twelfths and the range calculated at (b). Give the heights correct to 1 decimal place.

	high tide						*low tide*
time	17.10	18.10	19.10	20.10	21.10	22.10	23.10
height	8.1 m						

(6)

(d) The Albert Dock authorities gave the height at low tide at 23.10 on 22nd October 1988 as 3.0 m. By how much does your answer at (c) differ? *(1)*

(e) Here is the same table with the heights calculated using the mathematical formula:

	high tide						*low tide*
time	17.10	18.10	19.10	20.10	21.10	22.10	23.10
height	8.1 m	7.8 m	6.8 m	5.6 m	4.3 m	3.3 m	3.0 m

Calculate the differences at each hour between the heights given by the rule of twelfths and those given above. *(2)*

(f) Looking at your answers to (e), do you think sailors ought to carry a calculator and a maths textbook or continue to use the rule of twelfths in the future? Explain your answer. *(2)*

Total 15 marks

Answers to numerical questions

2 (a) 9 (b)(i) 1, 3, 6, 10 (ii) 15 (c) 45 (e) 165 (f) 35 (g) 44 cm
3 (a) 16 (b) 24 (e) 200 green, 240 blue
4 (d) 01.11.10 (f) 30
5 (b) 30 (c) 55 (e)(i) 91 (ii) 140 (iii) 204
7 (c) 4, 6, 8, 10, 12, 14, 16, 18, 20 22, 24, 26, 30 (f) 2, 28
8 (b) £160, £140, £88, £92 (c) 5 (d)(i) 4 (ii) 10%
9 (b) 48p (c) 84p
10 (a) 12 (b) 2 (c) 6 (d) 3 (e) 1 m (f) 66
11 (b) £12 (c) 60p (d) £12.60 (e) 63p (f) £11.97 (g) 3p (h) 84p
12 (a)(i) 10 (ii) 25 (iii) 9 (b) £105.77 (c) £1.06
13 (f) 2
14 (a)(i) £90 (ii) £40 (b) £235, £105
15 (b) 20%, 33⅓%, 50%, 66⅔, 80% (d) 10
16 (a)(i) 2.8 m, 4.2 m (ii) 1.4 m (iii) 50% (b)(i) 5.5 m, 4.9 m (ii) 0.6 m (iii) 10.9% (c)(i) 13.4 m, 11.6 m (ii) 1.8 m (iii) 13.4% (e)(i) 88.5%, 88.3%
17 (a) £2.56 (c) 11 (d)(i) £20.47 (g) $1\frac{255}{256}$
18 (b) 5m^2, 8m^2, 9m^2 (c) 5m^2 (d) 4
19 (a)(i) 2 cm^2 (ii) 1 cm^2 (iii) 3 cm^2 (b)(i) 4 cm^2 (ii) 12 cm^2 (d)(ii) 27 cm^2 (iii) 9
20 (b) 682 (c) £44.50 (d) £30,349 (e) £83.70
21 (a) £1.10 (b) 5 (c) 22p (d) 14p (f) 19.0751 (h) £20.98
22 (a)(i) 90 (ii) 7 ft 6 in (b)(i) 144 (ii) 12 ft (c)(i) 192 (ii) 16 ft (d) 3 ft (e) 5
23 (a) 22 cm (b)(i) 11 km (ii) 2¾ hrs (c)(i) 450 m (ii) 1½ hrs
25 (c) 2 (f) 2
26 (a) 1800, 3200, 5000 (b) 10,000 (c) 1734, 3174, 5046 (d) 9954 (e) A (f) 72
28 (c) 7 ft 6 in
31 (a) 120°, 80°, 80°, 80° (c) 2 (d)(i) 1.5 m^2 (ii) 1 m^2
32 (a) 3 ft
33 (j)(ii) 5 m
34 (a) 6 mins (b)(i) 13 m (ii) 6½ mins (c)(i) ½ min (ii) 8.3% (d)(i) 15 m (ii) 7½ mins (e)(i) 1½ mins (ii) 25%
35 (a)(i) 100 cm (ii) 43 cm (b) 86 cm
36 (a) 4 (b) 90° (c) 360° (d) 3 (e) 120° (f) 720°
37 (g) 25 km^2
38 (a) 0.00095 (b) 4737 (d) 23

39 (b)(i) £1.50 (ii) 95p (iii) 30p (iv) £1.80 (c) 60% (d) 50% (e)(i) £1.20 (ii) 76p (iii) 24p (iv) £1.44
40 (a) 7 hrs 50 mins (b) 16 hrs 38 mins
41 (a) 140, 120, 140, 180, 120, 180, 120, 180 (b) 30×30, 35×25, 40×20 (g) 140 cm
42 (d) 212°F (e) 25°C
43 (e) 4½ in (g) 6½
44 (f) 100%
46 (e) 24 m
47 (a) 1.4, 2.0, 2.6, 3.2, 4, 4.8, 5.8, 6.8, 7.8, 9
48 (a) 2ab (b) $(a \cos\theta + b \sin\theta)$, $(a \sin\theta + b \cos\theta)$ (c) 32 cm^2 (d) 14° (e) 8.25 cm, 3.88 cm (f) 32.0 cm^2 (g) $4 \text{ cm} \times 4 \text{ cm}$ (h) 32 cm^2 (i) 45° (j) 32 cm^2
49 (a) $2x$ (b) $2x - 8$ (c) $4x - 16$ (d) $4x - 24$ (e) $8x - 56$ (f) 7p (g) 8p
51 (a) 82% (b) 6%
52 (c) 13 yrs (d)(i) 12 (ii) 7 (iii) 0
54 (a) 33 (b) 11 (c)⅓ (d) ¼ (e) 4 (f) ⅕ (g) 4 (h) ¼
55 (d) 32% (e) 22%
56 (a) 14% (d)(i) 4 (ii) 1
57 (a) 1, ½ (b) 6 (d) 84p, £5.81, £4.72 (e) 87p, £6.15, £4.65
58 (c) 48.9%, 52.9% (d) 90%
59 (a)½ (b) ¼ (c) $\frac{1}{16}$
60 (a) £7,000 (b) £4,500 (e) £17,500 (g) £7,000 (h) £330,500
61 (c) 16 (d) 39
62 (a) 9 s (b) 1 min (c) 08.09 (d) 13 min (e) 08.28 (g) 8 min
64 (a) 6 (b) 21 (c) 6 (d) 12 (e) 42 (f) 5 (g) 91
65 (a) 3.86 m (b) 5.1 m (c) 7.7, 6.9, 5.6, 4.3, 3.5, 3.1 (d) 0.1 m (e) 0, 0.1, 0.1, 0, 0, 0.2, 0.1

British Library Cataloguing in Publication Data
Hudson, Sallie
Structured questions for GCSE mathematics.
Bk.1
1. Mathematics
I. Title
510

ISBN 0 340 50641 5

First published 1989

Typeset by Wearside Tradespools, Fulwell, Sunderland.
Printed in Great Britain for Hodder and Stoughton Educational, a division of Hodder and Stoughton Ltd, Mill Road, Dunton Green, Sevenoaks, Kent by Biddles Ltd, Guildford and Kings Lynn